# BLOOD BIOCHEMISTRY

CROOM HELM BIOLOGY IN MEDICINE SERIES

# Blood Biochemistry

N.J. Russell, G.M. Powell, J.G. Jones,
P.J. Winterburn and J.M. Basford

CROOM HELM
London & Canberra

© 1982 N.J. Russell, G.M. Powell, J.G. Jones, P.J. Winterburn and J.M. Basford
Croom Helm Ltd, Provident House, Burrell Row,
Beckenham, Kent BR3 1AT

British Library Cataloguing in Publication Data

Blood biochemistry. – (Croom Helm biology in
    medicine series)
    1. Blood – Chemistry
    I. Russell, N.J.
    621'.11      QP93

    ISBN 0-7099-0003-1
    ISBN 0-7099-0004-X Pbk

Set by Hope Services, Abingdon
Printed and bound in Great Britain
by Billing & Sons, Worcester

# CONTENTS

# PREFACE

The idea for this book arose from an integrated lecture course on the biochemistry of blood given to medical students in the second year of their pre-clinical studies. However, the material in that course has been expanded and it is intended that the book provide both the medical and non-medical reader with a concise and up-to-date account of the status of knowledge of the biochemistry of blood. A glance at the chapter titles shows how wide a field this covers, including many of the growth areas in biochemistry.

It is assumed that readers of the book will have a basic knowledge of biochemistry. A functional approach is adopted, and whenever possible the material is organised in terms of biochemical functions, although there are separate chapters on the white cell and the red cell. Because of the clinical importance of analysing blood components and assaying enzymes in the diagnosis of disease, chapters are included on the separation, preparation and measurement of blood components. In order to assist in bridging the gap between scientific studies and medicine, examples are included of changes associated with various diseases when the biochemistry is well understood. Attention also has been drawn to those areas where recent research in blood biochemistry has resulted in major advances in our understanding of fields having less obvious clinical relevance. It is hoped that these inclusions will stimulate interest and illustrate that such basic research is essential to the proper understanding of the functioning of blood in health and disease.

# ACKNOWLEDGEMENTS

We are grateful to Professor K.S. Dodgson for his support, and to Barbara Power and Jean Mitchell for typing the manuscript. We should also like to thank many of our colleagues for their helpful advice and in particular Professor G. Elder and Drs A. Cryer, C.G. Curtis, R. Eccles, R. John and M. Worwood.

# 1 INTRODUCTION

Blood has long held special significance for Man, assuming social and religious importance in many cultures. The ancient Greeks recognised that blood pulsated within vessels, and Aristotle taught that the heart was the centre of Man's body, its seat of life, from which all desires, feelings and emotions arose. The four chief fluids of the body were considered to be blood, yellow bile, black bile and phlegm, their balance determining the physical and mental qualities of the individual. A predominance of blood represented a sanguine [sic] temperament which, during the Middle Ages, was viewed with such special favour that, in pictorial representations, God was often given a ruddy complexion. William Harvey, during the seventeenth century, was the first to describe the circulation of the blood. He believed that blood was a uniform substance but, despite the inaccuracy of this belief, by recognising that blood circulated, Harvey paved the way for our present understanding of the complex composition and diverse functions of blood.

Vertebrates possess a high pressure blood system with a fast flow rate, which not only imposes some restrictions on certain components of blood, but facilitates a whole range of biochemical and physiological functions. For instance, in order that blood might flow at such a fast rate the red cell membrane must be flexible. A long-standing fascination about the biconcave shape of erythrocytes has led to fundamental discoveries of the molecular structure and function of the red cell membrane and membranes in general, as well as providing specific information on the mechanism of erythrocyte deformability and membrane alterations in ageing and disease. The rapid circulation of blood throughout the body provides an excellent 'postal system', allowing the speedy delivery of messages such as hormones from one organ to another, so co-ordinating the activities of different tissues.

The general functions of blood are in metabolism and its regulation, transport, osmotic balance and defence. The metabolic and transport roles of blood overlap to some extent, for instance in the carriage of oxygen and carbon dioxide. Studies on the structure and function of haemoglobin have been particularly fruitful areas of research, providing an enormous amount of information about such diverse fields as subunit interactions in multimeric proteins, protein biosynthesis and the structure of genes. This knowledge has led to biochemical explanations of several anaemias and molecular interpretations of some abnormal haemoglobins. Another aspect in which transport and metabolic functions are intimately linked is in the movement of metabolites. For example, glucose is utilised as an energy source by the cells within blood and is converted to pyruvate and lactate, which are returned to the liver for further metabolism. The important function of osmoregulation is yet another example of the blood functioning in both transport and metabolic capacities.

Blood plays an important part in the body's defence mechanisms. The immune response system is able to recognise foreign material within the body and a sequence of events is triggered that neutralises and destroys the foreign material. There are many recent advances in our knowledge of the biochemistry of these processes, such as the discovery only a few years ago of the human leucocyte antigen (HLA) system, the functions and significance of which are only just beginning to be appreciated. Efficient haemostasis is a pre-requisite of a high pressure circulation with a fast flow rate. Haemostasis minimises blood loss due to injury and prevents the entry of pathogens into the body. The practical importance of recent advances in some areas may not be immediately obvious; however, in other instances the significance is clearer. For example, once it had been realised that proteolysis was involved in blood coagulation, a better comprehension of the underlying enzymology opened up the possibility of improved anti-thrombotic drugs.

The complex composition of blood is not constant, but changes during stress, starvation, exercise and as the result of injury or disease. The fact that it is relatively simple to obtain blood samples and that blood composition reflects the overall state of the body has resulted in the widespread use of blood analysis as an aid to diagnosis. The analysis of various blood components, such as cells, proteins, metabolites and ions, and the assay of enzymes, provide useful information for the clinician. Proliferation of the number of parameters that can be measured and the pressure to process large numbers of samples rapidly has led to increasing automation. A highly competitive industry aimed at providing reliable biochemical tests that can be used as diagnostic aids has developed around the application of biochemical techniques to the separation and measurement of blood components.

# 2   RED BLOOD CELLS

**Introduction**

The red blood cell, or erythrocyte, possesses a most unusual cell shape. The discoid form, together with a distinctive red colouration, makes it one of the most easily recognised cell types in the body. It is also probably the most studied cell.

Adult blood contains $4.5-5.8 \times 10^{11}$ red blood cells per decilitre, with males generally having higher cell counts than females; these values correspond to haemoglobin contents of 14-15.5 g/dl blood. Using these figures it can be calculated that the haemoglobin concentration inside the red cell is extremely high, being approximately 35 per cent (w/v). A decrease in either the number of red cells or their haemoglobin content produces a condition known as anaemia (see Chapter 5).

## Cell Shape and Deformability

The discoid shape provides a large surface area to volume ratio and a thin cross-section, so that no haemoglobin molecule is more than 1 $\mu$m from the cell membrane. Consequently, the diffusion path for oxygen and carbon dioxide is short, which is an important factor in the major function of the red cell, namely gas carriage. Although red cells are approximately 8 $\mu$m in diameter, by deforming their shape they can pass through capillaries in the peripheral circulation that are considerably less than this size. The smallest pore size is encountered in the spleen. Blood entering the spleen from the Bilroth cords is forced through the endothelial cell basement membranes via slits 3.5 $\mu$m in diameter, which act as filters. During this process, although the shape of the red cell is deformed, the cell membrane is not stretched. The fact that red cells are able to deform rather than behave like rigid particles leads to blood behaving as a non-Newtonian fluid: its viscosity decreases as sheer forces are applied. If red cells were not flexible blood would be far more viscous; in fact it is doubtful that it could flow at all. This flexibility is a property of the cell membrane.

## Life Cycle

Red cells are made in bone marrow, where they arise from stem cells which differentiate to become reticulocytes. All these precursors are nucleated. Reticulocytes specialise in the production of two poly-peptides, $\alpha$- and $\beta$-globin, and protoporphyrin which, together with an iron atom, constitute normal adult haemoglobin. Inherited disorders in any of these syntheses can lead to anaemia. The reticulocytes begin a process of de-differentiation, losing their intracellular organelles such as mitochondria, nucleus, Golgi apparatus, ribosomes and endoplasmic reticulum. The mRNAs for $\alpha$- and $\beta$-globin are relatively long-lived so that globin synthesis can continue after the initial loss of the nucleus. By the time the mature red cell emerges into the blood stream it has lost the capacity to synthesise haemoglobin and retains only a limited but specialised metabolic capability. It is able to generate ATP for ion pumping and to synthesise specific compounds, such as 2,3-diphospho-glyceric acid and reduced glutathione, that are needed to maintain and regulate the function of haemoglobin. After approximately 120 days in the circulation the red cells are removed by the reticulo-endothelial system of the spleen and liver.

*Major Functions*

The major function of the red cell is the transport of oxygen from the lungs to the tissues. The amount of haemoglobin increases by seventy-fold the oxygen carrying capacity of the blood over that which can be carried in free solution. The presence of such a high concentration of haemoglobin inside the red cell creates a considerable osmotic pressure, and an osmotic balance is maintained by the transport of cations such as $Na^+$, $K^+$ and $Ca^{2+}$ (see Chapter 10). Ion transport is achieved by two types of ATP-dependent ion pumps in the cell membrane. If these pumps are inefficient or fail, the lack of osmotic regulation leads to an influx of water and a loss of the characteristic discoid shape. The red cell rounds up to form a spherocyte or may even form an echinocyte with spiny membrane projections on its surface. Red cells play a significant role in the control of blood pH: the $CO_2$ from tissues diffuses into the plasma and is taken up by red cells where it is converted into bicarbonate by the enzyme carbonic anhydrase. The bicarbonate, on passing back into the plasma, forms a major part of the buffering capacity of blood (see Chapter 10).

**Red Cell Membrane Composition and Structure**

The red cell membrane is the most studied of all cell membranes. There are several reasons for this. Erythrocytes can be obtained readily in large numbers, and as a cell type they are simple in possessing only one membrane. They can be broken gently in hypotonic media, their haemoglobin washed away, and the membranes resealed to form red cell 'ghosts'; these form a useful system in which to study membrane properties in the absence of cell metabolism. Another advantage in studying the erythrocyte is that its membrane has a relatively simple composition compared with many other cell membranes. However, like all other cell membranes it is composed largely of protein (49 per cent w/w) and lipid (43 per cent w/w), together with smaller amounts of carbohydrate (8 per cent w/w) present in glycoproteins and glycolipids. Although the total number of different proteins present in the membrane is not known, there are approximately twelve major ones, of which four possess covalently linked carbohydrate. These glycoproteins carry the bulk of the cell surface sialic acid. Many of the remaining proteins are present only in small amounts, but they nevertheless have important functions; for example the $Na^+$, $K^+$-ATPase, which maintains the membrane potential and controls the internal $Na^+$ and $K^+$ concentrations.

## Lipids and Glycolipids

The lipids of the red cell membrane are arranged as a bilayer of a mixture of cholesterol and phospholipid together with a small amount of glycolipid. This bilayer provides both a permeability barrier between the plasma and the inside of the red cell and a fluid milieu in which proteins and glycoproteins can function – as formalised in the now familiar fluid mosaic model of membrane structure. This general model for membrane structure was based largely on evidence from studies with red cell membranes. The membrane is not a rigid structure but is fluid with a viscosity corresponding to that of olive oil, within which both lipids and some proteins are able to diffuse rapidly in the plane of the membrane. In contrast to this lateral diffusion the rate of transfer of a lipid from one side of the membrane to the other, termed 'flip-flop', is many orders of magnitude slower. Proteins are believed not to 'flip-flop' at all. The various phospholipid classes (see Figure 2.1 for structures) are distributed asymmetrically in the membrane; most of the choline-containing phospholipids, phosphatidylcholine and sphingomyelin, are in the outer half of the lipid bilayer, whereas the amino-phospholipids, phosphatidylethanolamine and phosphatidylserine, are present mainly in the inner half of the bilayer. The rate of 'flip-flop' of phospholipids (half time = 20–30 minutes) is slow compared with

**Figure 2.1: Structures of Glycerophospholipids and Sphingomyelin**

(a)

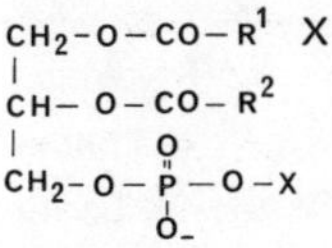

$CH_2-O-CO-R^1$  X  — CHOLINE IN PHOSPHATIDYLCHOLINE

$CH-O-CO-R^2$      — ETHANOLAMINE IN PHOSPHATIDYLETHANOLAMINE

$CH_2-O-\overset{O}{\underset{O_-}{P}}-O-X$  — SERINE IN PHOSPHATIDYLSERINE

$R_1$ AND $R_2$ ARE LONG CHAIN FATTY ACIDS $(C_{16} - C_{24})$ ON THE 1'- AND 2'-POSITIONS RESPECTIVELY.

GLYCEROPHOSPHOLIPID GENERAL STRUCTURE

(b)

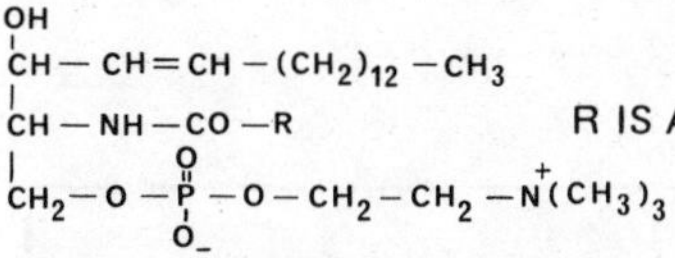

$\overset{OH}{CH}-CH=CH-(CH_2)_{12}-CH_3$

$CH-NH-CO-R$      R IS A LONG CHAIN FATTY ACID $(C_{16} - C_{24})$.

$CH_2-O-\overset{O}{\underset{O_-}{P}}-O-CH_2-CH_2-\overset{+}{N}(CH_3)_3$

SPHINGOMYELIN

their rapid movement in the plane of the membrane. However, in view of the 120 day lifespan of the red cell, 'flip-flop' would appear to be quite fast enough to break down lipid asymmetry. It is not understood how this asymmetry is created and maintained, neither is its biological significance appreciated. The distribution of cholesterol is still in some doubt, but evidence suggests that it is present mainly in the outer half of the bilayer.

The glycolipids, which more correctly should be termed glycosphingo-lipids (glycosyl ceramides), have 1–5 sugar residues attached to ceramide (acylsphingosine) and are present exclusively in the outer half of the bilayer. These glycolipids comprise a family of related structures, the major one of which in the human red cell membrane is globoside I (Figure 2.2). It is the carbohydrate residues of these glycolipids exposed on the outside of the cells which bear several antigenic determinants, such as those of the ABO system (see Chapter 8).

*Proteins and Glycoproteins*

The red cell membrane protein is composed of extrinsic (or peripheral) protein (approximately 40 per cent) and intrinsic (or integral) protein (approximately 60 per cent). Extrinsic protein is located on the inner face of the membrane and is made up mainly of two high molecular weight (200,000–250,000) proteins known collectively as *spectrin*, so named because it was first isolated from red cell 'ghosts'. Spectrin accounts for approximately 30 per cent of the total membrane protein. It exists as dimers and is associated with another extrinsic protein, which is smaller than spectrin and very similar in properties to the

**Figure 2.2: Structure of Globoside I**

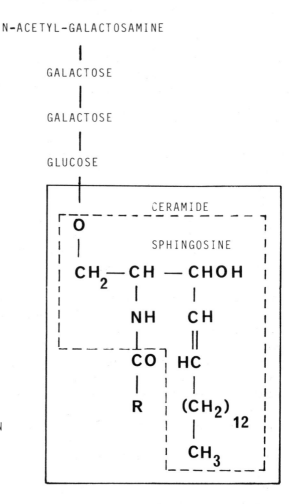

N-ACETYL-GALACTOSAMINE

GALACTOSE

GALACTOSE

GLUCOSE

CERAMIDE

SPHINGOSINE

R-ACYL CHAIN

muscle protein actin. Spectrin is usually isolated as a complex with several molecules of the actin-like protein, which under suitable ionic conditions *in vitro* (e.g. high $Ca^{2+}$ concentration) forms filaments. However, it is unlikely that the spectrin-actin complex is strictly analogous to muscle actomyosin, because spectrin is quite unlike muscle myosin and it has not been possible to show that the spectrin-actin complex is contractile. Other proteins may be loosely attached to the inner surface of the membrane, although the functional significance, if any, of this phenomenon is uncertain. Examples of such proteins are glyceraldehyde 3-phosphate dehydrogenase and possibly haemoglobin.

The separation of red cell membrane proteins by electrophoresis in polyacrylamide gels gives rise to a number of 'bands'. Most of the intrinsic protein migrates in 'band 3'; this is a diffuse heterogeneous band and it is not clear yet how many proteins or glycoproteins it contains. A major component is a glycoprotein, known as *glycophorin*, which is composed of 40 per cent protein and 60 per cent carbohydrate. The polypeptide portion comprises 131 amino acids whose sequence is known. Glycophorin is of particular interest because it is an example of an intrinsic protein that spans the membrane. The way in which the polypeptide is arranged across the membrane is not understood completely, but residues 73–92 have hydrophobic side-chains and if present in an α-helical conformation would be the correct length to span the hydrophobic core of the membrane phospholipid bilayer. The C-terminal portion of glycophorin is exposed on the inside of the red cell and bears no carbohydrate. The N-terminal portion, some 70 residues, on the outside of the red cell bears 16 short oligosaccharide chains, most of which terminate with a sialic acid residue. The sialic acid content of glycophorin accounts for 75 per cent of the sialic acid on the red cell surface. Some of its oligosaccharide chains, perhaps in conjunction with parts of the polypeptide chain, carry the MN antigenic activity (see p. 111). There are several red cell variants known that have abnormal amounts of surface sialic acid and which show altered MN antigenic expression. One rare type, known as En(a-), despite lacking glycophorin altogether, displays no clinical abnormalities. This observation casts doubt, therefore, on the importance of the role that glycophorin plays in the red cell.

Band 3 also contains polypeptides that have been implicated in the transport of ions, glucose and water across the red cell membrane. One of these proteins is the $Na^+$, $K^+$-ATPase; this transport protein is located in the inner half of the membrane, but it may well span the membrane. The $Ca^{2+}$-ATPase, whose function is to maintain the low intracellular $Ca^{2+}$ concentration, is another of the intrinsic proteins concerned with transmembrane movements of cations. A third transport protein is the anion exchange protein, which exchanges anions, such as $HCO_3^-$ for $Cl^-$, by a process of facilitated diffusion. This movement is extremely rapid – $10^{11}$ molecules of $HCO_3^-$ per red cell per second – corresponding to a turnover number for the carrier protein as great as that of the most efficient enzymes known. The anion exchange protein is one of the most abundant in the red cell membrane, comprising

30–35 per cent of the total protein. It has a monomer molecular weight of approximately 100,000 and spans the membrane with the polypeptide chain possibly recrossing the membrane several times. One model for anion transport proposes that the protein exists within the membrane as dimers, which could form specific channels for the passage of anions.

Although some minor proteins and glycoproteins remain uncharacterised, the organisation of the red cell membrane can now be visualised (Figure 2.3). This general picture enables us to explain certain of the fundamental characteristics of the erythrocyte, such as shape and flexibility. The intrinsic membrane proteins and glycoproteins are arranged in complexes that are large enough (6–8 nm diameter) to be seen by freeze-fracture electron microscopy. The presence of these intramembranous particles can be deduced also by various labelling techniques. On the cytoplasmic surface of red cell membranes the spectrin–actin network interacts with these intramembranous particles; the point of interaction is through another protein, sometimes called ankyrin, which links spectrin to band 3. This association restricts the movement of intramembranous particles and produces a co-ordinated matrix, which is probably responsible for maintaining the erythrocyte's biconcave shape, whilst at the same time retaining its flexibility and, therefore, ability to deform, for example, in narrow capillaries.

## Red Cell Ageing

As an erythrocyte ages it begins to lose its characteristic discocyte shape, the cell membrane loses lipid (all classes to an equal extent) and the cell density increases. This change in density is exploited experimentally in isolating old cells from blood by density gradient centrifugation. As the cell ages the membrane becomes leaky, the intracellular ATP concentration falls and the ion pumps function less efficiently. The question arises as to how these changes are linked to the loss of deformability in old cells and red cell destruction in the spleen.

One theory is based on the modification of spectrin. Experiments have shown that spectrin can be phosphorylated *in vitro* by protein kinases, and that only phosphorylated spectrin can interact with the contractile protein, myosin, from muscle. It is argued that if cellular ATP levels fall, spectrin will have fewer phosphate groups so leading to a reduction in the tension of the spectrin–actin network and a loss

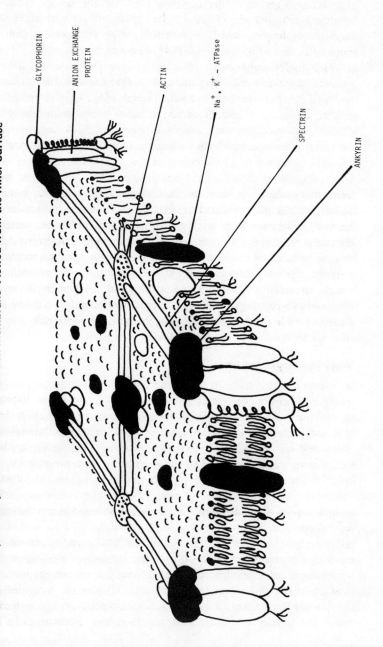

**Figure 2.3: Stylised Representation of the Red Cell Membrane Viewed from the Inner Surface**

GLYCOPHORIN

ANION EXCHANGE PROTEIN

ACTIN

$Na^+, K^+ - ATPase$

SPECTRIN

ANKYRIN

of the biconcave discoid shape. Defects in spectrin phosphorylation have been suggested as the common basis for the several types of hereditary spherocytosis. Erythrocytes from patients with myotonic muscular dystrophy, which is inherited as an autosomal dominant trait and is present in one in 6,000 of the population, have membranes in which the phosphorylation of a minor glycoprotein component is impaired. Such red cells have reduced mechanical resistance. Thus it may well be that membrane protein phosphorylation is important in red cell ageing and also in certain diseases where deformability is affected. However, this theory depends heavily on the analogy between the spectrin–actin complex and muscle actomyosin which, as pointed out earlier, may not be a valid comparison.

An alternative theory of red cell ageing, for which there is more convincing evidence, is based on the observed increase in $Ca^{2+}$ concentration that occurs in old erythrocytes. The $Ca^{2+}$ concentration rises to 0.5 mM, which is a level sufficient to activate the enzyme transglutaminase present in the cell membrane. Transglutaminases form crosslinks between proteins by creating iso-peptide bonds between $\epsilon$-amino groups of lysine residues and $\gamma$-carbonyl groups of glutamine residues; these bonds are similar to those formed in the crosslinking of fibrin (see p. 121) and the clotting of semen. Red cells with crosslinked membrane proteins are less flexible, and being unable to undergo reversible deformation are removed from the circulation by the spleen.

## Membrane Lipid Changes in Disease

Mature red cells synthesise very little lipid, but phosphatidylcholine, sphingomyelin and cholesterol in the outer half of the bilayer are able to exchange with phosphatidylcholine, sphingomyelin and cholesterol (but not cholesterol ester) in plasma lipoproteins. Red cell membrane cholesterol rapidly and freely exchanges with serum cholesterol, equilibrium being reached within eight to twelve hours. In contrast, phospholipids exchange more slowly, phosphatidylcholine being the fastest with <10 per cent exchange in twelve hours. The lipid composition of lipoproteins depends upon diet; such diet-induced changes in lipoprotein composition could lead to a change in the lipid content of red cell membranes. However, changes in the lipid composition of the membrane generally are minimal, with the amounts of cholesterol and phospholipid being maintained in approximately equimolar proportions. Changes in the cholesterol to phospholipid ratio or the nature of the phospholipids, particularly their fatty acid composition, would modify membrane fluidity and hence red cell flexibility. Alterations in the types of phospholipid or their fatty acid composition during disease are rarely seen; much more common is a change in the cholesterol to phospholipid ratio. This is brought about by the exchange of cholesterol between red cells and plasma. An important factor that affects this exchange is the activity of the plasma enzyme, lecithin-cholesterol acyl transferase (LCAT), which catalyses the following reaction:

$$\text{Cholesterol}_{plasma} + \text{Phosphatidylcholine}_{lipoprotein} \xrightarrow{\text{LCAT}}$$
$$\text{Cholesterol-ester}_{plasma} + \text{Lyso-phosphatidylcholine}_{lipoprotein}$$

in which the 2-position fatty acid (see Figure 2.1) of phosphatidylcholine is transferred to cholesterol. This enzyme is responsible for the formation of the majority of esterified cholesterol in plasma, and is inhibited by bile acids. Hepatocellular diseases, such as alcoholic cirrhosis, that lead to a rise in plasma bile acids (mainly chenodeoxycholic acid) result in an increase in plasma levels of unesterified cholesterol. The cholesterol equilibrates with red cell membrane cholesterol, thereby increasing the amount of cholesterol in the membrane. This has two effects: firstly, the cholesterol to phospholipid ratio is increased, which reduces membrane flexibility; secondly, the increased cholesterol content also raises the total lipid present in the membrane, expanding its surface area. An increase in total lipid content of only 8 per cent is enough to cause the formation of spherocytes. In alcoholic cirrhosis, where membrane cholesterol levels may rise by as much as 55 per cent, numerous spherocytes or even echinocytes are observed. These cells are removed from the circulation as a result of their altered size, shape and flexibility. Elevated levels of plasma bile acids (mainly cholic and deoxycholic acids) are observed in obstructive jaundice with similar consequences for the red cell membrane.

**Red Cell Metabolism**

The metabolism of the mature erythrocyte is organised for the performance of its major function — the delivery of $O_2$ to the tissues in adequate quantities. Because the erythrocyte has no nucleus, mitochondria, ribosomes, Golgi apparatus and endoplasmic reticulum, the enzyme systems associated with these organelles are absent. In consequence, the synthesis of DNA, RNA or protein is impossible, and the red cell is unable to replace denatured enzymes. The biochemical activities retained are those necessary for the efficient functioning of haemoglobin as a reversible $O_2/CO_2$ carrier and the preservation of cell integrity and shape. To these ends four compounds are required: (a) ATP – for the maintenance of membrane function, integrity and deformability; (b) 2,3-diphosphoglycerate (2,3-DPG) – to modulate the $O_2$ affinity of haemoglobin; (c) NADH – to maintain the haem in the Fe(II) state; (d) NADPH – to prevent haemoglobin denaturation.

## Glycolysis and 2,3-Diphosphoglycerate Production

The predominant metabolic fuel is glucose. This is converted by glycolysis, the erythrocyte's major metabolic pathway, into pyruvate and lactate which leave the cell in the absence of any citric acid cycle capability (Figure 2.4). Quantitatively, the most important tissue for the further metabolism of this pyruvate and lactate is the liver where they serve as gluconeogenic precursors. Thus a metabolic 'closed loop' is established with the red cell converting glucose to lactate and the liver resynthesising glucose from lactate. Commitment to the use of glucose by red cells imposes a considerable synthetic burden on the liver, particularly during fasting, since the red cells utilise 30–40 g glucose/day. The ATP provided by glycolysis is consumed mainly by ion pumps in the cell membrane. Failure to produce enough ATP results in an inability to maintain ionic balance leading to a change in shape as the cell swells, and an accumulation of $Ca^{2+}$ with a consequent loss of deformability (see p. 12). The NADH generated in glycolysis is a source of electrons for methaemoglobin reductase (see p. 19). Surplus NADH is reoxidised by the reduction of pyruvate to lactate. For this reason, under normal circumstances, the major product of glycolysis is lactate rather than pyruvate.

A metabolite unique to the red cell is 2,3-DPG. This is present at a concentration of 4–5 mM; that is almost equimolar with haemoglobin. 2,3-DPG binds in a 1:1 complex to the deoxy form of haemoglobin, so decreasing the affinity of haemoglobin for $O_2$ (see p. 32). The formation and breakdown of 2,3-DPG are catalysed by the two enzymes 2,3-diphosphoglycerate mutase and 2,3-diphosphoglycerate phosphatase, respectively (Figure 2.4). The flow of material through this shunt is important in two respects. Firstly, from a quantitative standpoint, 20–25 per cent of the material flowing through glycolysis passes via the shunt and therefore bypasses the ATP-forming reaction catalysed by 3-phosphoglycerate kinase. This decreases the yield of ATP from glucose. Secondly, the balance between the activities of the mutase and phosphatase determines the 2,3-DPG concentration and hence the modulation of haemoglobin affinity for $O_2$.

**Figure 2.4: Principal Metabolic Processes in the Erythrocyte**

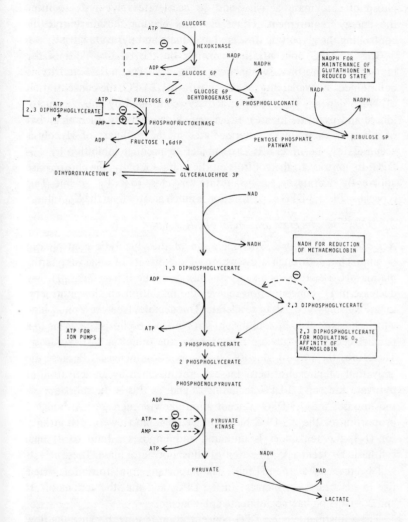

The key enzyme regulating the rate of glycolysis is phosphofructo-kinase. The activity of this enzyme is sensitive to changes in pH and the concentrations of several metabolites, of which AMP, ATP and 2,3-DPG appear to be the most important physiologically. AMP is an activator, whilst H⁺, ATP and 2,3-DPG are inhibitory. As in most cells the changes in AMP and ATP concentrations are inversely related — a decrease in ATP being mirrored by a rise in AMP and vice versa. Taken together these observations suggest that a demand for ATP increases phosphofructokinase activity and so accelerates glycolysis to meet the energy requirement. Other enzymes playing subsidiary roles in controlling the glycolytic flow are hexokinase and pyruvate kinase.

The 2,3-DPG concentration depends on the relative rates of the mutase and phosphatase reactions. The 2,3-DPG mutase activity is determined by availability of the substrate 1,3-DPG, the concentration of which depends on glycolytic activity (controlled by phosphofructo-kinase and pyruvate kinase). If pyruvate kinase activity is less than that of phosphofructokinase, intermediates in the mid-part of glycolysis accumulate. The mutase is also subject to product inhibition by 2,3-DPG. In contrast to the controls on the mutase activity, the phosphatase apparently operates at a constant rate very close to its $V_{max}$ value. This is because the 2,3-DPG concentration is much greater than the $K_m$ value.

### 2,3-Diphosphoglycerate as a Metabolic Regulator

*Acidosis and Alkalosis.* The operation of the glycolytic controls can be appreciated through a discussion of the effects of alkalosis. As the plasma pH rises there is a sympathetic rise in the intracellular pH. The decrease in $H^+$ concentration relieves the inhibition on phosphofructo-kinase causing glycolysis to accelerate. This depletes fructose 6-phosphate which, because it is in equilibrium with glucose 6-phosphate, leads to a parallel decrease in glucose 6-phosphate, the inhibitor of hexokinase. Therefore, the rate of glucose phosphorylation increases. Because the activation of phosphofructokinase is not matched by an activation of pyruvate kinase, 1,3-DPG accumulates and, as this is the substrate for the mutase, the 2,3-DPG concentration likewise increases. Although a proportion of the 2,3-DPG binds to haemoglobin (lowering its affinity for $O_2$), the remainder is unbound and imposes a limit to its own synthesis by feedback inhibition of phosphofructokinase. These effects will be reversed as the pH falls. The important point to note is that a rise in pH causes an increase in 2,3-DPG, and that the actions of pH and 2,3-DPG on haemoglobin are antagonistic.

The adjustment in 2,3-DPG concentration is slow, because the flow rate through the mutase reaction is only sufficient to renew the 2,3-DPG pool in the cell every 8–10 hours. Therefore, 2,3-DPG must be viewed as a long term modulator of oxygen binding to haemoglobin. In the short term the pH changes encountered on passing through metabolically active tissues are probably the most important. Transient alterations in pH modify haemoglobin affinity for $O_2$ and adjust the delivery of $O_2$ to tissues, but should the pH change be more lasting, then a compensatory adjustment in 2,3-DPG concentration is observed. The delayed response by 2,3-DPG can pose problems during treatment of acidosis or alkalosis. For example, if the blood pH of an acidotic patient is corrected quickly with a bicarbonate infusion, the Bohr effect (see p. 30) and low 2,3-DPG level will act together to increase the $O_2$ affinity of haemoglobin. This further impairs $O_2$ delivery to the tissues of a patient who is already ill. It is preferable, therefore, to correct the pH slowly enough for the 2,3-DPG level to readjust.

*Hypoxia.* There are a number of ways in which hypoxia can arise; for example, reduction in alveolar $pO_2$, anaemia or cardiovascular and pulmonary disease. The immediate consequence is that a greater proportion of the haemoglobin exists in the deoxy form. The binding of 2,3-DPG to deoxyhaemoglobin decreases the free 2,3-DPG concentration, which in turn relieves the inhibition on the mutase and phosphofructokinase. The increased activity of these two enzymes raises the 2,3-DPG concentration causing more oxyhaemoglobin to unload and so compensates for the lowered $O_2$ tension.

A related effect is observed when adjusting to the lower $O_2$ tensions present at high altitudes. Initially the hypoxia increases pH and the concentration of 2,3-DPG, leaving haemoglobin affinity for $O_2$ little changed. After a few days the body corrects the alkalosis returning the pH to normal, but 2,3-DPG levels remain high. The haemoglobin affinity is adjusted for the lower $O_2$ tensions. Over a period of a few weeks at altitude an increase in the number of red cells occurs.

### Maintenance of Native Haemoglobin

*Pentose Phosphate Pathway and Glutathione.* Haemoglobin denaturation is prevented by NADPH formed in the oxidative part of the pentose phosphate pathway. The flow of hexose units through this pathway is governed by the demand for NADPH, but is normally less than 10 per cent of the flow through glycolysis. Protein sulphydryl groups are prevented from being oxidised by the reduced form of glutathione. This tripeptide exists in two forms, oxidised and reduced — the former being a disulphide-bridged dimer while the latter has a free sulphydryl group (Figure 2.5). Glutathione is maintained in its reduced form by electrons transferred from NADPH in a reaction catalysed by glutathione reductase. In addition to preserving sulphydryl groups in proteins, reduced glutathione acts as a scavenger of free radicals and deals with hydrogen peroxide; these are highly reactive and are capable of denaturing proteins. The red cell synthesises glutathione from cysteine, glutamate and glycine, and so can replenish glutathione which has diffused out of the cell.

**Figure 2.5: Involvement of Glutathione in Erythrocyte Metabolism**

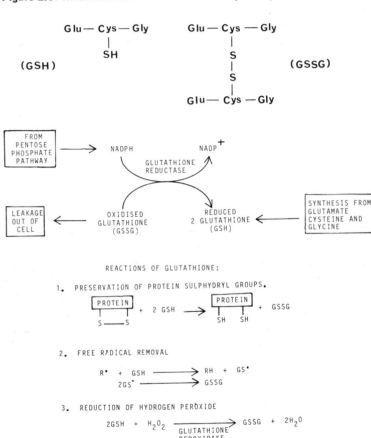

REACTIONS OF GLUTATHIONE:

1. PRESERVATION OF PROTEIN SULPHYDRYL GROUPS.

2. FREE RADICAL REMOVAL

3. REDUCTION OF HYDROGEN PEROXIDE

*Methaemoglobin.* Haemoglobin in which the iron atom is in the Fe(III) state is called methaemoglobin. Calculations suggest that up to 30 per cent of the haemoglobin in the normal individual is converted daily to methaemoglobin. This oxidation of haemoglobin occurs spontaneously, the causative agent usually being $O_2$ itself. When $O_2$ binds to haemoglobin an electron donated by the Fe(II) is shared between $O_2$ and the iron atom. Although as the $O_2$ dissociates this electron usually returns to the iron atom, occasionally the electron migrates to the $O_2$ creating a superoxide ion $(O_2^-)$. The iron is left in the III state. It is believed that iron oxidation is promoted by water entering the hydrophobic haem pocket. Interestingly, in accord with this explanation, certain abnormal haemoglobins have more accessible haem pockets and these forms exhibit increased rates of methaemoglobin and superoxide production (see p. 54).

The system for converting methaemoglobin back to haemoglobin is methaemoglobin reductase (Figure 2.6). The initial step is reduction of the iron atom of the haem groups by methaemoglobin reductase using NADH generated in glycolysis: the second stage is a non-enzymic transfer of an electron to methaemoglobin restoring the Fe(II) state.

**Figure 2.6: Regeneration of Haemoglobin from Methaemoglobin**

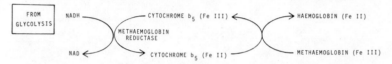

Should the cell's reductive capacity be exceeded, methaemoglobin accumulates and, being less stable than haemoglobin, dissociates into its haem and globin moieties. Once separated from the haem group the globin denatures, aggregates and precipitates inside the cell, usually bound to the cell membrane. These aggregates are visible using the light microscope and are known as inclusion bodies or Heinz bodies.

In some instances it may be beneficial to deliberately increase the amount of methaemoglobin. For example, one treatment for cyanide poisoning is inhalation of amyl nitrite which rapidly oxidises a proportion of the haemoglobin. The Fe(III) of methaemoglobin binds cyanide forming cyanomethaemoglobin which is non-toxic and is metabolised slowly. Thus, cyanide is effectively removed from the circulation and prevented from acting as a poison.

*Activated Oxygen.* The superoxide ion is a form of activated oxygen, a term which encompasses several highly reactive species all derived from superoxide (see Figure 6.5a). Certain of these activated forms are capable of denaturing proteins and reacting with membrane lipids leading to a loss of membrane functions. The removal of superoxide is achieved by a reductive process using NADPH generated in the pentose phosphate pathway. Superoxide dismutase catalyses the conversion of superoxide to $H_2O_2$ which is reduced by glutathione peroxidase (see Figure 6.5b). It follows from this that increased methaemoglobin formation leads to an increased demand for NADPH, to cope with the activated oxygen, and a stimulation of the pentose phosphate pathway.

## Further Reading

Gordesky, S.G. (1976) 'Phospholipid Asymmetry in the Human Erythrocyte Membrane', *Trends in Biochemical Sciences, 1,* 208–11

Gratzer, W.B. (1981) 'The Red Cell Membrane and its Cytoskeleton', *Biochemical Journal, 198,* 1–8

Harrison, R. and Lunt, G.G. (1980) *Biochemical Membranes: Their Structure and Function,* 2nd edn, Blackie, Glasgow

Peschle, C. (1980) 'Erythropoiesis', *Annual Review of Medicine, 31,* 303–14

Schrier, S.L. (1982) 'The Red Cell Membrane and its Abnormalities' in A.V. Hoffbrand (ed.), *Recent Advances in Haematology*, Ch. 4, Churchill Livingstone, Edinburgh

### The Function of Haemoglobin

A resting adult extracts about 5 ml of oxygen (measured at STP) from every 100 ml of circulating blood. During exercise, this figure can increase almost three-fold. Since the solubility of oxygen in plasma is only 0.3 ml/100 ml, it is not surprising that blood contains about 15 g/100 ml of an oxygen transporter. This transporter is the protein haemoglobin and is contained entirely within the erythrocyte.

The properties of haemoglobin are such that it is fully saturated with oxygen in arterial blood where the oxygen tension in plasma is about 100 mm of Hg. Since 1 g of haemoglobin can bind 1.4 ml of oxygen, the oxygen content of arterial blood is about 20 ml/100 ml of blood. At rest, the haemoglobin releases 23 per cent of its oxygen load and the oxygen tension in plasma drops to 40 mm of Hg. During exercise a further 40 per cent of the original oxygen load is released from the haemoglobin with a relatively small drop in oxygen tension to 20 mm of Hg. Any additional requirement for oxygen is met largely by an increase in cardiac output. This remarkable capacity of haemoglobin to bind and release oxygen is illustrated graphically in Figure 3.1 where it is compared with properties predicted for a protein in simple equilibrium with oxygen as follows:

$$P + O_2 \rightleftharpoons P(O_2)$$ 
(3.1)

The binding of oxygen by haemoglobin follows a sigmoidal, rather than the conventional hyperbolic, curve. This is important for the physiological function of haemoglobin – releasing oxygen to tissues (see Figure 3.1). Another important consequence of the sigmoidal binding curve is that it can be displaced horizontally with very little change in oxygen-loading but a large effect on the amount released to the tissues (see Figure 3.2). This kind of displacement can be produced by competitive inhibitors of oxygen binding and there are three such inhibitors which function as physiological modulators of oxygen transport.

(1) *Hydrogen ions.* The association of haemoglobin with oxygen is pH-dependent. At pH values greater than 6.5 the affinity of haemoglobin

for oxygen increases with increasing pH. This effect is known as the alkaline Bohr effect. A corollary to this is the release of hydrogen ions when haemoglobin binds oxygen.

**Figure 3.1: Whole Blood–Oxygen Dissociation Curve (A) and a Theoretical Curve (B) Predicted from Equation (3.1) and Assuming the Same Oxygen Content as Whole Blood Under Arterial Conditions.** The length of the arrow represents the extra oxygen obtained from the sigmoidal curve when the venous partial pressure of oxygen is 20 mm Hg.

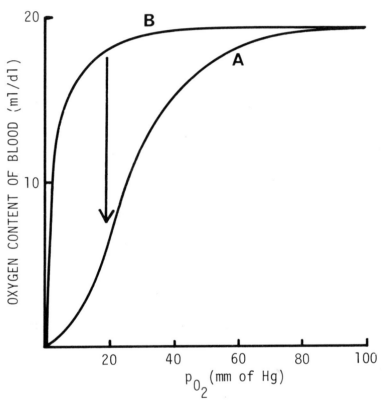

(2) *Carbon dioxide.* Oxygen binding to haemoglobin is inhibited by carbon dioxide for two reasons. Firstly, an increase in concentration produces a decrease in pH with its concomitant effect on oxygen binding (the Bohr effect; see above). Secondly, carbon dioxide itself binds to the N-terminal amino groups of haemoglobin and reduces the affinity of the protein for oxygen. The effect of

both carbon dioxide and hydrogen ions is to decrease the affinity of haemoglobin for oxygen, *not as it is loaded in the lung, but as it is released to the tissues.*

(3) *2,3-Diphosphoglycerate (2,3-DPG).* The affinity of haemoglobin for oxygen, as measured *in vitro*, is much greater than that of whole blood due to the presence of a competitive inhibitor in the erythrocyte. This inhibitor is 2,3-DPG which acts as an important modulator of the function of haemoglobin *in vivo*.

**Figure 3.2: Whole Blood–Oxygen Dissociation Curve (A) Compared to that in the Presence of a Hypothetical Competitive Inhibitor (B).** The length of the arrow represents the extra oxygen obtained from the inhibited sample at a venous partial pressure of oxygen of 20 mm Hg.

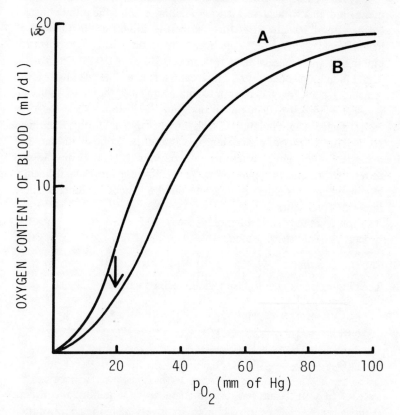

These three physiological modulators of haemoglobin–oxygen affinity do not act independently. The effect of each inhibitor depends on the concentration of the others. They act together to provide a situation in which haemoglobin can release the maximum portion of its oxygen load at an oxygen tension high enough to ensure an adequate rate of diffusion of the gas into respiring tissues.

## Quantitative Descriptions of the Properties of Haemoglobin

The haemoglobin molecule has a number of intriguing properties which are reflected in the mode of binding of oxygen to haemoglobin in whole blood. The association curve is sigmoidal and the binding is inhibited *competitively* by molecules which are not structural analogues of oxygen. These properties are characteristic of multi-subunit (oligomeric) proteins with more than one binding site for the primary ligand (oxygen) and *separate sites* for competitive inhibitors. In the case of haemoglobin the sigmoidal binding curve is the result of *co-operative* binding of four successive oxygen molecules to a tetrameric protein. The binding of one oxygen molecule increases the affinity of the remaining sites for the next molecule. Similarly the third molecule is easier to pick up than the second and the loading of the fourth is easier again. The binding of such competitive inhibitors has been termed an *allosteric* interaction, because these compounds are not isosteric (of the same structure) with oxygen. It is now conventional to refer to proteins with these properties as allosteric proteins and haemoglobin became a model on which to build our understanding of these important molecules.

Three methods are commonly used to describe the co-operative properties of allosteric proteins.

### The Hill Equation

The binding of oxygen to haemoglobin can be written:

$$
\begin{aligned}
Hb &+ O_2 \rightleftharpoons Hb(O_2) \\
Hb(O_2) &+ O_2 \rightleftharpoons Hb(O_2)_2 \\
Hb(O_2)_2 &+ O_2 \rightleftharpoons Hb(O_2)_3 \\
Hb(O_2)_3 &+ O_2 \rightleftharpoons Hb(O_2)_4
\end{aligned}
\tag{3.2}
$$

Each of these reactions has its own characteristic equilibrium constant. If, however, the degree of co-operativity is very high, the concentration of intermediate species will be low compared to those of Hb and $Hb(O_2)_4$; and under these circumstances the oxygenation reactions (3.2) can be described by one equilibrium constant:

$$K = \frac{[\text{Hb}]\,[O_2]^4}{[\text{Hb}(O_2)_4]} \tag{3.3}$$

The proportion ($Y$) of binding sites of haemoglobin occupied by oxygen is then given by the expression:

$$\log[Y/(1-Y)] = 4\log[O_2] - \log K \tag{3.4}$$

A plot of $\log[Y/(1-Y)]$ against $\log[O_2]$ will be linear with a slope of four. Equation (3.4) describes adequately the binding of oxygen to haemoglobin in the region of 10 per cent to 90 per cent saturation. However, the assumption has been made in devising expression (3.3) that the concentration of intermediate species is negligible; since they are not, the slope of the line is not four but about three. This slope is termed the Hill constant or coefficient ($n$) and is an empirical measure of the degree of co-operativity. A value of one would signify that all oxygen molecules bind equally well, that is no co-operativity and a hyperbolic binding curve, whereas the maximum value of four would reflect the maximum degree of co-operativity.

## The Monod, Wyman and Changeaux (MWC) model

This description is based on a molecular theory of ligand binding to allosteric proteins. It assumes *two* conformational states of haemoglobin, the tense ($T$) state having a lower affinity for oxygen than the relaxed ($R$) state. The $T$- and $R$-states exist in free equilibrium with one another; the $T$-state predominates in the absence of oxygen and relaxes to the $R$-state on oxygenation. The following equilibria are envisaged:

$$4\,O_2 + \qquad \rightleftharpoons \qquad + 4\,O_2 \tag{3.5}$$

$$3\,O_2 + \qquad \rightleftharpoons \qquad + 3\,O_2$$

$$2\,O_2 + \qquad \rightleftharpoons \qquad + 2\,O_2$$

$$O_2 + \quad \boxed{\begin{array}{cc} O_2 & O_2 \\ O_2 & \end{array}} \quad \rightleftharpoons \quad \text{(R-state)} \quad + \quad O_2$$

$$\boxed{\begin{array}{cc} O_2 & O_2 \\ O_2 & O_2 \end{array}} \quad \rightleftharpoons \quad \text{(R-state)}$$

*T*-state             *R*-state

The association curve (Figure 3.1) is described by three constants, $L$, $K_T$ and $K_R$, which refer to the equilibria (3.6), (3.7) and (3.8) respectively.

$$T_O \rightleftharpoons R_O \text{ (absence of oxygen)} \tag{3.6}$$

$$T + nO_2 \rightleftharpoons T(O_2)_n \tag{3.7}$$

$$R + nO_2 \rightleftharpoons R(O_2)_n \tag{3.8}$$

These constants can be used to describe the binding of oxygen according to the following equation:

$$Y = \frac{L\,C\,\alpha\,(1 + C\alpha)^3 + (1 + \alpha)^3}{L\,(1 + C\alpha)^4 + (1 + \alpha)^4} \tag{3.9}$$

where   $\alpha = O_2/K_R$   and   $C = K_R/K_T$

This model and equation agrees with experimental measurements from zero to 100 per cent saturation.

*The Koshland, Némethy and Filmer (KNF) model*

This description is based again on a molecular theory but allows for the existence of more than two conformational states of the protein. The following equilibria are envisaged:

$$\boxed{\begin{array}{cc} & \\ & \end{array}} \rightleftharpoons \boxed{\begin{array}{cc} O_2 & \\ & \end{array}} \rightleftharpoons \boxed{\begin{array}{cc} O_2 & O_2 \\ & \end{array}} \rightleftharpoons \text{...} \rightleftharpoons \text{...} \tag{3.10}$$

Each step in the sequence has its own equilibrium constant ($K_1$, $K_2$, $K_3$ and $K_4$), because the binding of each oxygen molecule is assumed to improve the binding of the next. The binding curve is described by the following equation:

$$Y = \frac{K_1[O_2] + 2K_1K_2[O_2]^2 + 3K_1K_2K_3[O_2]^3 + 4K_1K_2K_3K_4[O_2]^4}{4(1 + K_1[O_2] + K_1K_2[O_2]^2 + K_1K_2K_3[O_2]^3 + K_1K_2K_3K_4[O_2]^4)}$$

(3.11)

Of the three models, the Hill equation is the least accurate but is that most commonly employed in a clinical context. This is because it is the simplest with only two constants, and is adequate over the range of saturations which are relevant *in vivo*. The value of $K$ (equation 3.3) is, however, handled more conveniently as $n\sqrt{K}$, which is termed the $P_{50}$, and is the tension of oxygen in solution required to occupy half the available binding sites (50 per cent saturation). The Hill equation has no basis in molecular theory and usually is replaced by the simple molecular model (MWC) when discussing the relationship between the structure and function of haemoglobin.


### The Structure of Haemoglobin

Haemoglobin is roughly spherical with a diameter of 5.5 nm and an overall molecular weight of 54,400. It is a tetrameric protein which consists of two different kinds of subunits ($\alpha_2\beta_2$); the $\alpha$- and $\beta$-subunits contain 146 and 141 amino acid residues respectively. The overall shapes of the $\alpha$- and $\beta$-subunits are remarkably similar and each carries a haem group. The subunits are arranged in such a way that the four haem groups are equidistant from one another and each occupies a cleft in the surface of the protein. A model of oxygenated haemoglobin is shown in Figure 3.3.

Six electron-donating groups can be accommodated around a positively charged Fe(II) ion. In a molecule of haem, four of these groups are the nitrogen atoms of the protoporphyrin (see p. 42 for structure). Two of these nitrogen atoms have lost a proton so that the overall complex is uncharged. In haemoglobin the fifth position is occupied by a nitrogen atom of a histidine residue in the polypeptide chain. This residue is sometimes referred to as the proximal histidine (His 87$\alpha$ or His 92$\beta$). The sixth co-ordination position of the iron can either be empty or occupied by an oxygen molecule.

**Figure 3.3: A Low Resolution Model of Oxyhaemoglobin Based on X-ray Crystallographic Measurements.** The arrow indicates the position of the glutamate residue which is replaced by valine in haemoglobin S (sickle cell haemoglobin).

Source: This photograph was a gift from Dr M. Perutz.

Haemoglobin will form crystals in a concentrated solution because each molecule has the same ordered structure. If the crystals are formed under nitrogen, they will crack when exposed to oxygen. This is a startling demonstration that molecules of oxyhaemoglobin have a different ordered structure from that of deoxyhaemoglobin. In the nomenclature of the MWC model, the deoxyhaemoglobin is the *T*-state and the oxyhaemoglobin the *R*-state. The molecular details of the difference between the two structures have been determined by X-ray diffraction analysis of both types of crystals. In general, both molecules have a high content of α-helical structure and are divided into seven helical regions termed A–G. This nomenclature serves to simplify the positioning of amino acids in the subunits for comparisons between haemoglobins from different species. For example, the proximal histidine is found as the eighth amino acid residue in helix F (His F8; i.e. His 87α or His 92β) in both α- and β-subunits. The *differences* between the structures of oxyhaemoglobin and deoxyhaemoglobin allow a structural interpretation of co-operative and allosteric interactions.

## Oxyhaemoglobin – R-state

In the tetramer, each $\alpha$-subunit is attached to both $\beta$-subunits through two areas of contact.

(1) The contact between $\alpha_1$- and $\beta_1$-subunits ($\alpha_1\beta_1$ contact) is the more extensive and involves some 34 non-polar amino acid side chains. These contacts form very strong hydrophobic interactions between the $\alpha_1$- and $\beta_1$-subunits; identical interactions bind the $\alpha_2$- and $\beta_2$-subunits.

(2) The $\alpha_1\beta_2$ (and $\alpha_2\beta_1$) contact regions connect the two dimers; each contact is of a hydrophobic nature and involves some 19 amino acid side chains.

In addition, each subunit contributes about 18 amino acid residues to hydrophobic interactions with the haem group, and there is the covalent bond between the Fe(II) atom and the proximal histidine residue (see above).

## Deoxyhaemoglobin – T-state

When oxygen is removed the major movement is in the $\alpha_1\beta_2$ contact with the two $\beta$-subunits rotating apart by about 0.7 nm. The new $T$-state is stabilised by extra hydrophobic and electrostatic bonds involving the C-terminal regions of each subunit. These have the following structures:

$\alpha$ ---------------- Tyrosine – Arginine $\ominus$
$\qquad\qquad\qquad\qquad\qquad\qquad \oplus$

$\beta$ ---------------- Tyrosine – Histidine $\ominus$
$\qquad\qquad\qquad\qquad\qquad\qquad \oplus$

Each of the penultimate tyrosine residues is held in a new hydrophobic contact between two helical segments of the same subunit. This anchors each bivalent C-terminal residue in a position favourable for electrostatic bond formation with other side chains. The $T$-state is further stabilised by the binding of 2,3-diphosphoglycerate (2,3-DPG), which has no affinity for the $R$-state. This is made possible by the separation of the $\beta$-subunits when oxygen dissociates from the $R$-state, allowing the 2,3-DPG to enter the cavity and form an electrostatic bridge between the subunits.

### The T → R Transition

In the T-state, the Fe(II) atom is in a high spin state (four unpaired electrons), but it changes to a low-spin state (no unpaired electrons) when oxygen binds. This loss of spin is probably accompanied by a decrease in radius of the iron atom allowing it to move into the centre of the porphyrin ring. The movement of the iron displaces the histidine to which it is attached and hence the remainder of the polypeptide chain.

## Structure–Function Relationships in Haemoglobin

The movements within and between subunits form the basis of a satisfactory explanation of the important physiological properties of haemoglobin.

### The Sigmoidal Binding Curve

When oxygen binds to one subunit, the resulting movement of the polypeptide expels the penultimate tyrosine from its hydrophobic contact and breaks the electrostatic bonds involving the adjacent C-terminal amino acid. The equilibrium between the T- and R-states will then move in favour of the R-state, fewer bonds being left to stabilise the T-state. The equilibria (3.5), therefore, will be displaced further to the right by progressive oxygenation. Thus, each successive oxygen molecule will encounter a higher proportion of vacant binding sites in the R-state; hence the affinity increases and the binding curve is sigmoidal. These movements are summarised in Figure 3.4.

### The Bohr Effect

The electrostatic bonds of the T-state involve the positive charges of the N-terminal residues of $\alpha$-subunits and the C-terminal histidines of $\beta$-subunits. When these bonds are broken on oxygenation, the p$K_a$ values of the participating groups decrease and protons are released (0.7 protons per molecule of oxygen). This effect will be significant only with the groups mentioned which have p$K_a$ values of about 7.4 (plasma pH). Thus, the binding of oxygen to haemoglobin can be written as:

$$Hb + 4 O_2 \rightleftharpoons Hb(O_2)_4 + 2.8 H^+ \tag{3.12}$$

**Figure 3.4: Diagrammatic Representation of the Quaternary Structure of Deoxygenated (A) and Fully Oxygenated (B) Haemoglobin.** The electrostatic bonds, which are broken on oxygenation, are indicated.

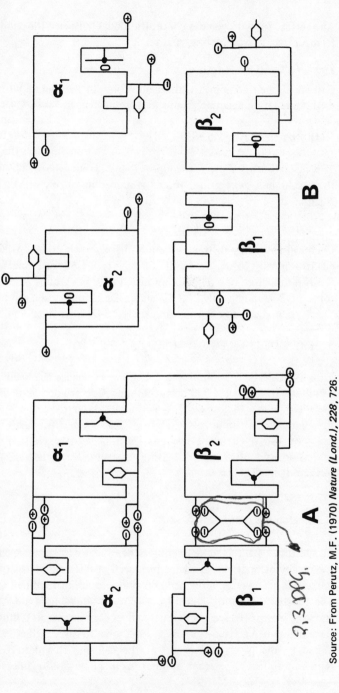

Source: From Perutz, M.F. (1970) *Nature (Lond.). 228*, 726.

An increase in hydrogen ion concentration will displace the equilibrium to the left and favour deoxygenation.

## The Effect of Carbon Dioxide

Carbon dioxide combines with haemoglobin in the $T$-state by binding to those N-terminal amino groups that are uncharged.

$$Hb.NH_2 + CO_2 \rightarrow Hb.NH.COOH \tag{3.13}$$

The formation of these carbamino groups favours the stabilisation of the $T$-state and thereby promotes the unloading of oxygen at respiring tissues.

## The Effect of 2,3-Diphosphoglycerate

Chemical and X-ray diffraction studies have shown that 2,3-DPG binds electrostatically to the $\beta$-subunits through lysine 82, histidine 143 and the N-terminal groups. The juxtaposition of these groups is favourable for 2,3-DPG binding only in the $T$-state. The closing of the gap between the $\beta$-subunits during oxygenation expels 2,3-DPG from its binding site.

$$Hb(2,3\text{-DPG}) + 4\,O_2 \rightleftharpoons Hb(O_2)_4 + 2,3\text{-DPG} \tag{3.14}$$

Thus 2,3-DPG will inhibit oxygen binding and cause the haemoglobin to unload its oxygen. Fetal haemoglobin differs from that of the adult in being an $\alpha_2\gamma_2$ tetramer. The $\gamma$-subunits differ from the $\beta$-subunits in a number of amino acid residues, but notably the histidine 143 $\beta$ is replaced by a serine residue. This change decreases the affinity of haemoglobin for 2,3-DPG giving fetal blood a higher affinity for oxygen than adult blood.

## Haemoglobin Variants

In addition to the $\alpha$- and $\beta$-subunits described earlier, there are four genetically-distinct globin chains produced at different stages of development. The corresponding haemoglobins are summarised in Table 3.1. The remaining haemoglobins (4 per cent) of adult blood are $A_{1a}$, $A_{1b}$ and $A_{1c}$, which are glycosylated derivatives of normal adult haemoglobin (haemoglobin A). Haemoglobin $A_{1c}$ is known to carry a glucose residue linked to the N-terminal group of the $\beta$-chain. This derivative may represent about 20 per cent of adult haemoglobin during uncontrolled diabetes. There is also a derivative of fetal haemoglobin (F) formed by acetylation of the N-terminal groups of the γ-chains. This comprises about 20 per cent of fetal haemoglobin at birth. In addition, there are about 270 known abnormal genetic variants of haemoglobin which have been named after the discoverer or, more usually, after the town where they were discovered. These variants may be due to either a single amino acid substitution or a difference in the number of amino acid residues in one of the globin chains.

**Table 3.1: Normal Genetic Variants of Human Haemoglobin**

| Haemoglobin | Subunit structure | Normal environment | % of haemoglobin in adult blood |
|---|---|---|---|
| A | $\alpha_2 \beta_2$ | Adult blood | 92 |
| $A_2$ | $\alpha_2 \delta_2$ | Adult blood | 3 |
| F | $\alpha_2 \gamma_2$ | Fetal blood | <1 |
| Gower 1 | $\delta_2 \epsilon_2$ | Early embryonic blood | nil |
| Gower 2 | $\alpha_2 \epsilon_2$ | Early embryonic blood | nil |
| Portland | $\delta_2 \gamma_2$ | Early embryonic blood | nil |

*Single Amino Acid Substitutions* (238 variants)

Over half of these are physiologically normal because the substitution occurs in positions on the surface of the molecule. The remaining variants can be grouped according to their physical properties.

(1) *Unstable Haemoglobins* (64 variants). The stability of the tetramer depends on the internal siting of non-polar groups, the integrity of the haem to globin contacts and the extensive $\alpha_1 \beta_1$ contacts mentioned earlier. Amino acid substitution in any of these regions gives rise to an unstable molecule and a congenital Heinz-body haemolytic anaemia (see Chapter 5).

(2) *Haemoglobins with Altered Oxygen Affinity* (27 variants). The affinity of haemoglobin for oxygen depends on the integrity of the $\alpha_1 \beta_2$ contacts, the 2,3-DPG binding site, haem binding regions and the C-terminal portion of each subunit. Substitutions within any of these regions give rise to blood with an altered oxygen affinity. The majority (24) of these variants have an increased affinity for oxygen and are usually associated with a mild polycythaemia (i.e. increased red cell number). The remaining three variants show a decreased affinity for oxygen and two are associated with anaemia. Some of these variants are also unstable.

(3) *Haemoglobins with Stabilised Fe(III) – Methaemoglobins* (five

**Table 3.2: Examples of Single Amino Acid Substitutions in the C-terminal Region of Abnormal Haemoglobins**

| Haemoglobin | A | Rainier | Bethesda |
|---|---|---|---|
| Base sequence | -AAG-UAU-CAC-UAA-[a] | -AAG-UGU-CAC-UAA-[a] | -AAG-GAU-CAC-UAA-[a] |
| Amino acid sequence | -Lys-Tyr-His | -Lys-Cys-His | -Lys-Asp-His |
| Oxygen affinity | normal | increased | increased |
| Bohr effect | normal | decreased | decreased |

Note: a. UAA is one of the signals for termination of protein synthesis.

variants). There are five variants with a substitution in the micro-environment of the haem group. The abnormal amino acid is either tyrosine (four variants) or glutamate. The presence of the electronegative oxygen atom stabilises the iron in the Fe(III) state and renders its reduction more difficult (see Chapter 2). The resulting methaemoglobin gives the blood a characteristic brown appearance and reduces its oxygen-carrying capacity. Examples of how some of these variants arise and how the properties of the haemoglobins are affected are shown in Table 3.2.

All the variants described above are rare, being confined to families. In contrast, haemoglobin C and haemoglobin S (sickle cell haemoglobin) are common among certain racial groups (see p. 52). Apart from being relatively common, both variants are interesting because they involve a single amino acid replacement on the surface of the protein molecule. All other such variants are physiologically normal. However, the replacement in haemoglobin S of Glu.$6\beta$ with a hydrophobic valine residue creates symmetrical faces on the $\beta$-subunits that allow molecules in the $T$-state to aggregate to form long fibres. These aggregates precipitate and cause the characteristic sickling of the erythrocytes. The solubility of haemoglobin in the $R$-state, which predominates at high oxygen tensions, is unaffected. One approach to the chemical treatment of sickle cell disease is the use of compounds which stabilise the $R$-state and increase the affinity of the haemoglobin for oxygen. Initially, isocyanate was used, which reacts with the $\alpha$-amino groups of N-terminal valine residues on the $\alpha$- and $\beta$-subunits to give a stable carbamoyl derivative. (This is similar to the physiological reaction between $CO_2$ and haemoglobin which yields a carbamino derivative.) Unfortunately, the adverse side-effects of isocyanate have limited its use. A more recent alternative based on the same principle is acetylation of HbS with acetyl-3,5-dibromosalicylic acid, a derivative of aspirin, which is sufficiently lipophilic to penetrate the red cell. Thus, it is possible that the deleterious effects of sickling can be avoided at the expense of a small decrease in the availability of oxygen to tissues.

*Molecular Size Variants* (32 variants)

Another group of physiologically abnormal haemoglobins is that in which one of the subunits is abnormal in length. These abnormalities may have arisen for one of three reasons.

(1) *Termination Errors.* (a) There are four variants with elongated

α-subunits due to a *single base substitution* in the termination codon of the α-gene. In haemoglobin Constant Spring the first new amino acid, attached to the original C-terminal arginine, is glutamine. The codon for glutamine (CAA) could have arisen from the original termination signal (UAA). (b) There is one variant with an elongated α-subunit and two with elongated β-subunits; all three variants are formed as a result of *frame shift mutation* (addition or deletion of a single base), which modifies the termination codon. The following scheme describes a possible cause of the elongated α-subunit in haemoglobin Wayne:

Normal base sequence      -A CC-UCC-A A A -U A C -CGU-U A A -GC.-
Normal amino acid
sequence                  -Thr-Ser-Lys-Tyr-Arg

The deletion of an adenine base (A) from the lysine codon gives:

Base sequence             -A CC-UCC-A A U -A CC-GUU-A A G -C..-
Amino acid sequence       -Thr-Ser-Asn-Thr-Val-Lys-

and the chain would continue to the next termination sequence.

(2) *Sequence Repetition.* In haemoglobin Grady, the α-subunit is lengthened due to the repetition of the tripeptide sequence (Glu-Phe-Thr) of positions 116–18 inclusive.

(3) *Subunit Fusions.* The genes coding for β- and δ-globin are adjacent on the chromosome. Three variants of haemoglobin have arisen, apparently, by fusion of that part of the gene coding for the C-terminal portion of a normal β-chain (60–90 residues) with the part of the gene coding for the N-terminal portion (50–80 residues) of a normal δ-chain. This hybrid subunit is found in haemoglobin Lepore in place of the β-subunit in the tetramer.

## Further Reading

Bunn, H.F., Forget, B.G. and Ranney, H.M. (1977) *Human Haemoglobin*, W.B. Saunders & Co., Philadelphia
–– (1977) *Haemoglobinopathies*, W.B. Saunders & Co., Philadelphia
Perutz, M.F. (1978) 'Haemoglobin Structure and Respiratory Transport', *Scientific American, 239*, 92–125
–– (1979) 'Regulation of Oxygen Affinity of Haemoglobin: Influence of Structure of the Globin on the Haem Iron', *Annual Reviews of Biochemistry, 48*, 327–86

# 4   HAEMOGLOBIN: SYNTHESIS AND DEGRADATION

Red cells have a limited life span. Their constant breakdown and synthesis imposes a requirement for the degradation and synthesis of haemoglobin. The degradative and biosynthetic pathways are subject to numerous controls. Much is known about them because they are particularly amenable to detailed investigations. Also, in many instances, the clinical consequences of malfunction can be explained.

## Haemoglobin degradation

At the end of their life span, the red cells are removed from the circulation and the haemoglobin which they contain is degraded. The main sites of degradation are the reticulo-endothelial cells of spleen, bone marrow and liver. Senescent cells also may lyse within the blood stream. The haemoglobin released is bound to haptoglobin, transported to the liver and taken up by the hepatic parenchymal cells. If the haptoglobin-binding capacity is exceeded, the free haemoglobin may be absorbed and degraded by the renal epithelial cells. Thus, in intravenous lysis, both the liver and the kidney can be important sites of haemoglobin breakdown.

Although some of the details of the events of haemoglobin breakdown are still uncertain there is no doubt about the principal features of those events:

(a) the constituents of the haemoglobin molecule are eventually separated;

(b) the globin is degraded by intracellular proteolysis to its constituent amino acids which are returned to the amino acid pool; and

(c) the haem is metabolised to produce the bile pigments which are excreted in the bile and the iron is returned to the iron stores (see Chapter 5).

### The Degradation of Haem

Scission of the haem moiety occurs between the symmetrically substituted pyrroles (Figure 4.1). An asymmetrical molecule is produced

**Figure 4.1: The Degradation of Haem to Form Bilirubin**

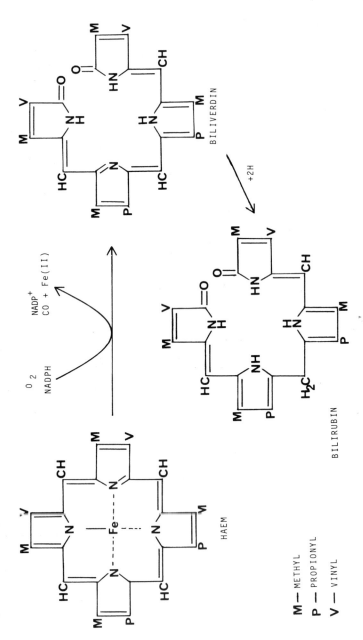

M — METHYL
P — PROPIONYL
V — VINYL

consisting of four pyrrole rings joined by three carbon bridges. The sequence of the four methyl, two vinyl and two propionate side chains is identical with that found in the protoporphyrin ring of haem after removal of carbon at the α-methene bridge. There has been much controversy over the question of whether haem catabolism is enzymic or non-enzymic. However, it is well known that several enzyme systems can convert haem to the bile pigments. Haem oxygenase, an endoplasmic reticulum enzyme, has received most attention because its activity is greatest in the spleen, bone marrow and liver, the sites of haem degradation. The details of that degradation are not clear. It has been postulated that the haem is bound to cytochrome p450, located in the membrane of the endoplasmic reticulum, and a reactive oxygen radical is produced which attacks the haem at its α-bridge. In any event, it is known that the process involves loss of the α-carbon atom as CO, as well as oxidation of the two adjacent carbon atoms. The product, biliverdin, subsequently is reduced to bilirubin by biliverdin reductase (see Figure 4.1), an enzyme which is found in most mammalian tissues. Bilirubin is virtually insoluble in aqueous solutions because of its highly lipophilic nature which, in turn, is due to the strong intramolecular hydrogen bonding. It is also highly toxic. Biliverdin, on the other hand, is a harmless compound and is the biliary excretory product of haem in birds, amphibians and reptiles. It is difficult to rationalise the formation of the toxic bilirubin in mammals but a possible explanation may be that bilirubin, unlike biliverdin, can cross the placenta and thus can be removed from the mammalian fetus.

## Fate of Bilirubin

The fate of bilirubin can be described under four main headings:

(1) *Transport in the Blood.* Bilirubin is transported in the plasma tightly bound to albumin (see Chapter 11). Several drugs are known to compete with bilirubin for albumin binding sites and so displace bilirubin (see p. 141).

(2) *Hepatic Uptake and Transport in the Hepatocyte.* At the sinusoidal plasma membrane, bilirubin is detached from albumin and transported into the hepatocyte. The transport process appears to be carrier-mediated. In the hepatic cytosol there are several non-specific proteins (aryl-glutathione transferases) which bind bilirubin reversibly, for example the Y (ligandin) and Z proteins. The precise roles which these proteins fulfil remain matters for conjecture. The significance of binding may be that it allows sequestration of the pigment in

non-toxic form. It has been claimed that binding improves pigment solubility and is an essential prelude to the transfer of bilirubin from the plasma membrane to the endoplasmic reticulum where it is metabolised.

(3) *Conjugation.* The appearance in bile of the highly lipophilic bilirubin is dependent on conjugation. Conjugation transforms bilirubin, providing it with physico-chemical characteristics which make it suitable for excretion via the bile. Glucuronic acid appears to be the dominant conjugating group and while in most animals the monoconjugates predominate, in man diconjugates are usually present in greater amounts. The main site of synthesis of the mono-

**Figure 4.2: Formation of Bilirubin Monoglucuronide**

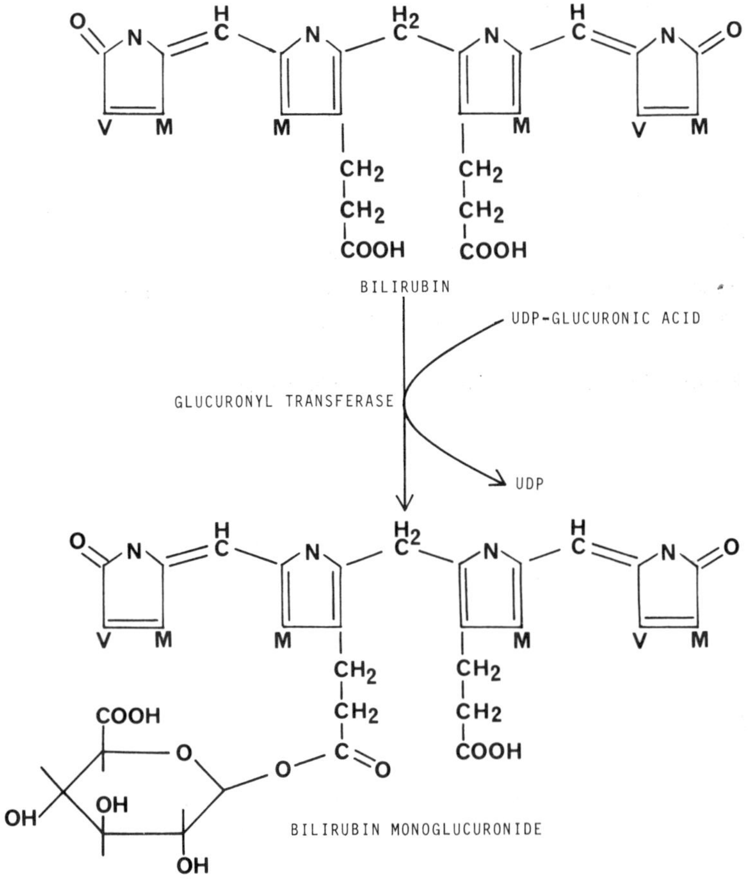

glucuronide is the endoplasmic reticulum; UDP-glucuronic acid is the glucuronyl donor (see Figure 4.2). Within the hepatocyte, bilirubin is present almost entirely as the monoconjugate and it has been suggested that conversion to the diconjugate may occur at the surface membrane of the liver cell.

(4) *Secretion into Bile.* Precisely how conjugated bilirubin is secreted into the bile is not known. It is secreted against a concentration gradient and a carrier-mediated active transport system probably is involved.

## Synthesis of Haemoglobin

The synthesis of haemoglobin requires three components:

    (a) a porphyrin (protoporphyrin IX),
    (b) globin,
    (c) iron.

Protoporphyrin IX is synthesised in the developing erythrocyte from glycine and succinyl-CoA (Figure 4.3). When the formation of the porphyrin is complete Fe(II) is added to produce haem. The haem enters the pocket of a globin polypeptide chain where it is protected from oxidation to haemin (Fe(III) state).

**Figure 4.3: The Synthesis of Haemoglobin in Outline**

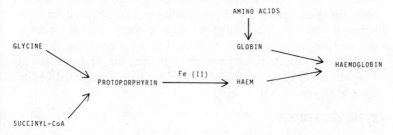

## *Synthesis of Protoporphyrin*

The biosynthetic pathway has been deduced largely from isotopic tracer studies in reticulocyte preparations. Glycine and succinyl-CoA contribute atoms to the final porphyrin molecule as shown in Figure 4.4, and the biosynthetic sequence is shown in Figure 4.5. The first,

**Figure 4.4: The Structure of Protoporphyrin IX**

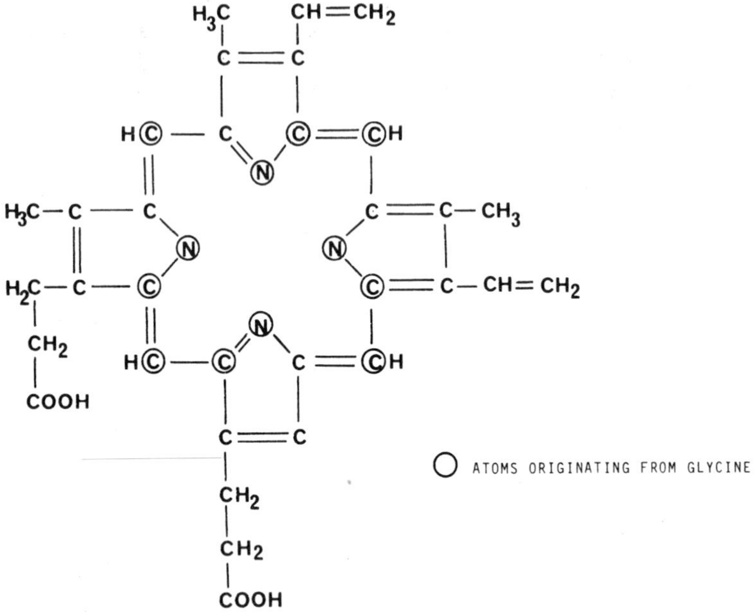

○ ATOMS ORIGINATING FROM GLYCINE

and rate-limiting, step of porphyrin biosynthesis, catalysed by the enzyme 5-aminolevulinic acid synthetase, involves the condensation of glycine with succinyl-CoA to form succinyl glycine. The subsequent loss of $CO_2$ results in the formation of 5-aminolevulinic acid, two molecules of which condense to form the pyrrole derivative, porphobilinogen. Four molecules of porphobilinogen are required to form a porphyrin ring. Both a synthetase and a co-synthetase are needed for this reaction. It is important to note that the intervention of the co-synthetase produces the asymmetrical isomer, uroporphyrinogen III. The side chains of uroporphyrinogen III are successively modified and protoporphyrin IX is produced.

## Synthesis of Globin

Globin synthesis in reticulocytes provides a valuable system for studying protein biosynthesis in general, because haemoglobin represents over 90 per cent of the protein made by the reticulocytes. The reason for this seems to lie in the restricted range of mRNA produced.

The genes for α- and β-globin are carried on separate chromosomes, numbers 16 and 11 respectively, and there are probably two copies of the α-gene in most populations. The γ- and δ-globin genes, coding for the globin chains which precede β-globin during development, are located also on chromosome 11. The genes for δ- and β-globin in particular are closely linked and a deletion mutation bridging the junction of these two genes is responsible for the creation, by gene fusion, of Hb Lepore (see p. 36).

**Figure 4.5: Biosynthesis of Protoporphyrin IX**

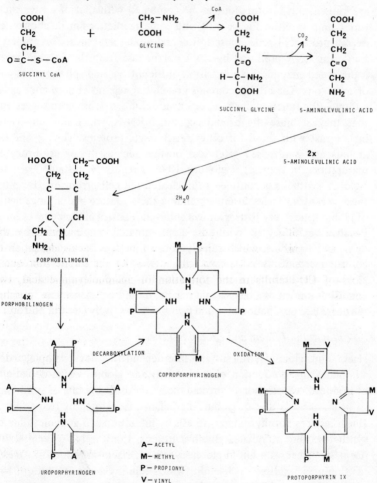

One of the most startling recent discoveries in molecular biology is that many genes in eukaryotes contain intervening nucleotide sequences that do not code for amino acids; these intervening sequences are known as introns and the coding regions as exons. The human $\beta$-globin gene contains two introns, a short intron between codons 30 and 31 and a larger intron of some 900 base pairs between codons 104 and 105 (see Figure 4.6). Introns also occur in $\gamma$- and $\delta$-globin genes and probably in the $\alpha$-globin genes. The functions of introns in eukaryote genes are a matter of speculation, but may be involved in the maturation of mRNA.

Since the complete sequence of $\beta$-globin mRNA and $\beta$-globin itself are known, this information can be combined with that of gene structure (summarised in Figure 4.6), and some predictions made about the overall process of gene transcription and globin synthesis. The primary transcript RNA produced by RNA polymerase from the $\beta$-globin gene is processed enzymically by excision of the introns and splicing together of the exons. The mRNA contains non-coding regions at both ends and, although the function of these regions is unknown, it has been suggested that they stabilise the $\alpha$- and $\beta$-globin mRNAs, which are long-lived in contrast with those for other reticulocyte proteins. Thus $\alpha$- and $\beta$-globins are synthesised after loss of the nucleus during erythrocyte maturation. In contrast, $\delta$-globin mRNA is relatively unstable, so that $\delta$-globin synthesis is confined to nucleated red cell precursors. Despite these recent advances in understanding the structure and arrangement of globin genes, very little is known about the control of their expression. Because reticulocytes synthesise approximately equimolar amounts of $\alpha$- and $\beta$-globins, which rapidly form dimers, it seems certain that control mechanisms exist, especially in view of the duplication of $\alpha$-globin genes. If the balanced synthesis of $\alpha$- and $\beta$-globin is disturbed a condition known as thalassaemia results; $\alpha$- and $\beta$-thalassaemia refer to the decreased production of $\alpha$- and $\beta$-globin respectively (see Chapter 5).

### Co-ordination of Haem and Globin Synthesis

Haem and globin synthesis are interdependent. For example, iron stimulates the production of both haem and globin, whilst depletion of iron leads to cessation of protein biosynthesis. If the rate of synthesis of haem exceeds that of globin, the balance is restored by two effects. Haem reduces porphyrin biosynthesis by inhibition of 5-aminolevulinate synthetase, and stimulates globin synthesis. Conversely, a decrease in haem levels causes a fall in globin synthesis. Haem produced in excess of globin is oxidised to haemin, which reinforces the inhibition, by haem, of its own biosynthesis. Haemin is required for the formation of the initiation complex, the obligatory step in the assembly of mRNA and ribosomes, so that in the absence of haemin, protein synthesis does

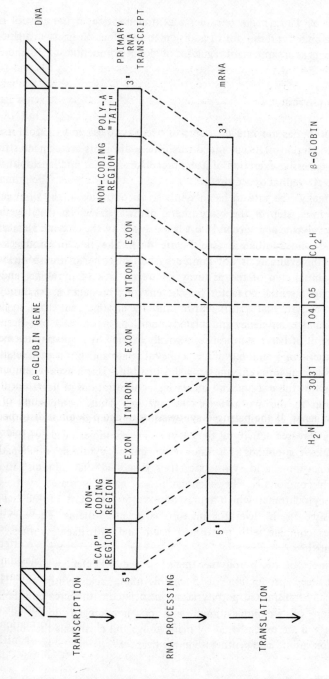

**Figure 4.6: Structure of the β-Globin Gene**

not occur. Thus, haem occupies a central position in the control of haemoglobin synthesis for in addition to its effect on haem synthesis, it also acts as an important regulator of globin production.

## The Porphyrias

The porphyrias are rare disorders of haem synthesis in which there is over-production of haem precursors. These disorders are characterised by the excessive excretion of haem precursors in urine and faeces. They may be congenital or acquired.

Normally, the rate of haem synthesis is controlled at the first, and rate-limiting, step of the biosynthetic pathway in which the reaction between glycine and succinyl-CoA is catalysed by the enzyme 5-amino-levulinic acid synthetase (see Figure 4.5). The rate of production of 5-aminolevulinate is not limited by lack of either of the substrates in the initial step of the pathway. However, lack of pyridoxal phosphate, an essential co-factor for the enzyme, produces a decrease in enzyme activity that is sufficient to cause the anaemia that characterises pyridoxamine deficiency and related conditions.

Control of haem synthesis is normally exerted by regulating the rate of synthesis of 5-aminolevulinic acid synthetase and this is particularly effective because the enzyme has a short half-life. Haem usually controls synthesis of the enzyme and increasing concentrations of haem cause a reduction in the synthesis of the enzyme. Thus, compounds that increase haem catabolism or that inhibit haem synthesis result second-arily in increased activity of the enzyme. For example, a wide variety of lipophilic compounds are known to induce synthesis of hepatic 5-aminolevulinic acid synthetase. It is suggested that administration of such compounds, like phenobarbitone, cause increased synthesis of apocytochromes which represent over 50 per cent of the haem synthesised in the liver for microsomal haemoproteins. These cyto-chromes combine with preformed haem and thus haem synthesis is stimulated.

In most of the porphyrias there is a genetic defect which results in enhanced production of 5-aminolevulinic acid synthetase. The genetically-transmitted porphyrias are associated with increased urinary excretion of 5-aminolevulinate and porphobilinogen during acute attacks and are exacerbated by many drugs and chemicals that induce 5-aminolevulinic acid synthetase in the liver.

## Normal Excretion of Haem Precursors

Normal urine contains small amounts of porphyrins (coproporphyrin and uroporphyrin) and porphobilinogen (see Figure 4.5). The characteristic spectroscopic properties of the porphyrins led to their early detection in normal and pathological urines. Urine containing excessive amounts of porphyrins may be variously coloured from pink to wine-red or may show a reddish-purple fluorescence in ultra-violet light.

The amounts of porphyrin excreted daily in normal urine are in the microgram range, while 0.4-2 mg of porphobilinogen is excreted each day. Similarly, normal faeces contain only small amounts of porphyrins (coproporphyrin, less than 20 $\mu$g/g dry weight; protoporphyrin, less than 30 $\mu$g/g dry weight and uroporphyrin, 10–40 $\mu$g in 24 hours). The prefixes uro- and copro- indicate the medium from which each of these compounds was originally isolated, but it is now known that both may be excreted in either urine or faeces.

## Routes of Excretion of Haem Precursors

5-Aminolevulinate and porphobilinogen are excreted exclusively in the urine, but the biliary route is the major one for porphyrin excretion. Both coproporphyrin and protoporphyrin are excreted largely in the bile and both of these porphyrins undergo an enterohepatic circulation during which some protoporphyrin may be re-utilised for haem synthesis.

The distribution of porphyrin excretion between urine and bile depends on the type of porphyrin and porphyrinogen as well as on the excretory functions of the liver. In general, the more hydrophilic a porphyrin the greater its urinary excretion. This is exemplified by the fact that the water-soluble 8-carboxy-uroporphyrins and uroporphyrinogens are excreted in urine but the 2-carboxy-porphyrins and porphyrinogens are excreted in bile.

Liver malfunction may result in distortion of urinary and biliary excretion. For example, liver disease leads to the diversion of biliary coproporphyrins to urine. Cholestasis also causes redistribution of porphyrin thus favouring urinary excretion.

## Abnormal Excretion of the Porphyrins and their Precursors

Conditions in which excessive amounts of porphyrins and their precursors are excreted in urine or faeces or both may be categorised as either congenital or acquired. The porphyrias may be further categorised by division into two main classes according to whether excessive production of porphryin occurs in the bone marrow (erythropoietic porphyrias) or, more commonly, in the liver. The latter group frequently shows increased production of 5-aminolevulinate and porphobilinogen.

As previously stated, the porphyrias are usually congenital, often exhibiting a marked familial incidence, and are included among the so-called 'inborn errors of metabolism'. The clinical manifestations are of two types – either skin lesions due to the photosensitisation of the skin or acute attacks of severe abdominal pain, often accompanied by neurological or psychiatric symptoms and frequently precipitated by drugs such as barbiturates.

The biochemical manifestations of the porphyrias are best illustrated by reference to the types of hepatic porphyrias which have been described. For example, *acute intermittent porphyria*, the most common of the familial hepatic porphyrias and an inherited autosomal dominant disorder is characterised by elevated hepatic levels of 5-aminolevulinic acid synthetase and depressed hepatic levels of uroporphyrinogen 1 synthetase. Consequently, excessive amounts of aminolevulinic acid and porphobilinogen are excreted in urine; on standing, the urine becomes red owing to the spontaneous formation of porphobilin and porphyrins from porphobilinogen. *Congenital cutaneous hepatic porphyria* is also an autosomal dominant disorder. The major clinical symptom is cutaneous photosensitivity exacerbated by wavelengths of light in the region in which porphyrins absorb intensely (about 400 nm). There is increased urinary excretion of aminolevulinate, porphobilinogen and porphyrins. Another autosomal dominant disorder of haem metabolism is *hereditary coproporphyria* and in this condition the outstanding observation is the increased excretion of coproporphyrin in the faeces. In addition, the levels of aminolevulinate, porphobilinogen and coproporphyrin in the urine may be elevated. This disorder may be provoked by drugs and other chemicals that induce 5-aminolevulinate synthetase in the liver. Cutaneous sensitivity is also the major clinical symptom in *acquired hepatic porphyria*. This disease appears not to be inherited, although the possibility of an underlying genetic abnormality exacerbated by hepatotoxic agents, like alcohol, has been suggested. The porphyrin excreted in excessive amounts in urine is mainly uroporphyrin. In this condition the urinary levels of aminolevulinate and porphobilinogen are usually normal.

## Further Reading

Elder, G.H. (1980) 'Haem Synthesis and Breakdown' in A. Jacobs and M. Worwood (eds), *Iron in Biochemistry and Medicine II*, Ch. 7, Academic Press, New York

Lawn, R.M., Efstratiadis, A., O'Connell, C. and Maniastis, T. (1980) 'The Nucleotide Sequence of the Human β-globin Gene', *Cell, 21*, 647–51

Marks, G.S. (1981) 'The Effects of Chemicals on Hepatic Heme Biosynthesis', *Trends in Pharmacological Sciences, 2*, 59–61

Ostrow, J.D. (1974) 'Bilirubin and Jaundice' in F.F. Becker (ed.), *The Liver. Normal and Abnormal Functions. Part A*, Ch. 11, Marcel Dekker, Inc., New York

Shaeffer, J.R., McDonald, M.J. and Bunn, H.F. (1981) 'Assembly of Normal and Abnormal Human Haemoglobins', *Trends in Biochemical Sciences, 6*, 158–61

Weatherall, D.J. and Clegg, J.B. (1979) 'Recent Developments in the Molecular Genetics of Human Haemoglobin', *Cell, 16*, 467–79

# 5   THE ANAEMIAS

## Introduction

The term anaemia applies to a group of diseases in which the blood may contain fewer erythrocytes than usual or cells with a decreased haemoglobin content. These conditions may be caused by a number of factors: thus blood loss, accelerated cell destruction or impaired erythropoiesis may produce anaemia either independently or in concert. However, the consequences for the patient are a decrease in oxygen-carrying capacity or an impaired oxygen delivery to the tissues, which are manifested by the symptoms of fatigue, faintness, headache and lack of concentration. Although the main symptoms are almost universal in anaemic patients, there is no simple unifying origin for the diseases.

**Table 5.1: A Classification of the Origins of Anaemia**

---

1. Decreased number of erythrocytes due to shortened life-span:

    (a) abnormal haemoglobin — e.g. affecting solubility or stability;
    (b) abnormal membrane proteins — e.g. defects of cell shape;
    (c) abnormal enzyme — e.g. glucose-6-phosphate dehydrogenase;
    (d) infection — e.g. malaria;
    (e) immune disorders — e.g. haemolytic disease of the newborn.

2. Decreased production of erythrocytes:

    (a) impaired haemoglobin synthesis — e.g. iron-deficiency;
    (b) defect in stem cell maturation — e.g. vitamin $B_{12}$ deficiency.

3. Abnormal red cells.

---

In fact the causes are many and varied. Three broad categories can be identified (see Table 5.1), and within these categories there are inherited and acquired conditions. Most inherited anaemias, like other congenital diseases, are extremely rare; but two such anaemias, glucose 6-phosphate dehydrogenase deficiency and sickle cell anaemia (see below), are often encountered. In general though, the acquired anaemias are the more common, and iron-deficiency anaemia is the most prevalent form throughout the world.

Cell size and shape are useful indicators of the anaemia type. Size is a parameter easily measured with modern electronic cell counters and microscopic examination of a blood smear permits assessment of shape and identification of other cell types, for example reticulocytes. On this basis anaemias are classed as *micro-*, *normo-* or *macrocytic*, according to whether the erythrocytes are respectively smaller, normal or larger in size, while terms such as *spherocytes* (spherical) or *echinocytes* (spiky) describe their shape.

The body has several compensatory mechanisms, which co-operate in delaying the onset of anaemia. Erythrocyte metabolism adjusts to the impaired oxygen delivery by increasing the concentration of 2,3-diphosphoglycerate. This modulates the oxygen affinity of haemoglobin by binding to the deoxy form thus displacing the equilibrium so that oxygen dissociates more readily at the low oxygen tensions in the tissues. This enables the residual haemoglobin to function more efficiently. In addition, erythropoietin (see p. 70) is released in response to the poor delivery of oxygen and accelerates the rate of erythrocyte maturation in the bone marrow. Similarly, in a haemolytic anaemia the bone marrow responds by making more cells; sometimes these are released prematurely leading to increased numbers of reticulocytes in the circulation. When these compensatory mechanisms become inadequate, then the symptoms of anaemia appear.

## Haemolytic Anaemias

These are characterised by an increased rate of erythrocyte destruction. In milder forms the erythrocyte life-span is shortened from its normal 120 days to 15–20 days, and in severe cases the cells survive for only a few days. However, the bone marrow can increase by up to eight-fold its production of red cells so that symptoms appear only when the erythrocyte life-span is reduced to about 20 days.

Haemolytic anaemia is often accompanied by jaundice. Normally haemoglobin, released by erythrocyte lysis, is degraded. The haem group is converted to bile pigments, which are conjugated in the liver; but in a haemolytic disease the capacity of the liver is exceeded and unconjugated bile pigments accumulate in the blood and tissues. When haemolysis is severe more haemoglobin is liberated than can be bound by haptoglobin, the specific salvage protein for haemoglobin in the plasma. The haemoglobin is, therefore, excreted in the urine and the resulting dark colour of the urine is the explanation for the term 'blackwater fever'.

*Inherited Haemolytic Anaemias*

*Erythrocyte Membrane Defects.* The commonest form of congenital haemolytic anaemia in whites is *hereditary spherocytosis*. The incidence is approximately one in 5,000. The condition is characterised by the presence of spherocytes, which are osmotically more fragile and which are destroyed in the spleen. The molecular basis is uncertain, but the abnormality probably arises from a defect in the organisation of the spectrin–actin network. The membrane may be leaky to $Na^+$ ions causing increased activity of the $Na^+$, $K^+$-ATPase to maintain ionic gradients, which in turn increases the demand for ATP. This is reflected in a higher glycolytic rate than normal. As the cells pass through the spleen small membrane fragments are lost; the cells become rounded to compensate for the decreased surface area and eventually lyse.

*Abnormal Haemoglobins.* Among the large number of haemoglobin variants known there are forms in which the solubility, stability or oxygen affinity are affected (cf. p. 33). Although most of these abnormal haemoglobins are extremely rare, a few are common.

   *Sickle cell anaemia* is a disease associated with the haemoglobin form known as HbS. The gene for HbS occurs with a high frequency in most of Africa, Greece and the Indian subcontinent, and those people who originate from these regions. It is estimated that 9–10 per cent of American blacks are affected by sickle cell disease, whereas the incidence is about 30 per cent in some areas of Africa. The homozygous condition shows the severe form of the disease, whereas the heterozygote (one copy of the gene) has the sickle cell trait and usually experiences no clinical problems. About 10–15 per cent of homozygous children do not survive beyond two years of age. Electrophoretic analysis of the haemoglobin in homozygotes reveals that up to 80 per cent is HbS with the remainder being the fetal form, HbF. The heterozygous individual produces approximately equal amounts of normal and abnormal haemoglobins.

   In HbS the glutamate residue at the $\beta6$ position is substituted by a valine residue. This reduces the solubility of the deoxy ($T$-state) form although the oxyhaemoglobin solubility is unaffected. On deoxygenation HbS precipitates in long filaments that cause distortion of the erythrocyte membrane into the characteristic sickle shape. Initially, this transition is reversible particularly if the circulation, and hence reoxygenation, is rapid. However, sickle cells tend to block capillaries causing the oxygen tension to fall further. This aggravates the sickling, rendering the tissues anaerobic, and leads to pain and tissue damage due to anoxia.

By this time, the cells are permanently sickled, more fragile and liable to haemolysis. In consequence, affected persons must avoid situations where oxygen tensions are low such as travelling in unpressurised aircraft or climbing mountains.

In addition to HbS, there are three other common variants (HbC, HbD and HbE) which also have decreased solubility in the $T$-state, although this is not as marked as in HbS; consequently the anaemia is milder. All these variants are common in certain parts of the world: HbC is confined mainly to West Africa, HbD to Iran and India and HbE to south-east Asia. In all the variants there is a single amino acid substitution in the $\beta$-chain. The occurrence of all these abnormal haemoglobins, including HbS, correlates geographically with the incidence of malaria. The protection from malaria offered by the presence of the abnormal haemoglobins offers a possible explanation for the common occurrence of abnormal haemoglobin genes. However, although the heterozygous state provides some protection against a common parasitic disease, this is at the expense of a severely disadvantaged homozygote.

The *thalassaemias* are a group of diseases in which the structure of the globin subunits is normal but there is an imbalanced synthesis of $\alpha$- and $\beta$-chains. In $\alpha$- and $\beta$-thalassaemia the synthesis of the $\alpha$- and $\beta$-chains are depressed. In $\beta$-thalassaemia, the more common form, inhibition of $\beta$-chain synthesis leads to an accumulation of $\alpha$-globin, whereas the converse happens in $\alpha$-thalassaemia. Homozygotes, presenting with the disease *thalassaemia major*, produce very little functional haemoglobin and severe anaemia develops at an early age. In the heterozygous condition the disease (*thalassaemia minor*) takes a much milder course. $\beta$-Thalassaemia is a common disease in some parts of the world, notably the Middle East, south-east Asia and throughout the Mediterranean, and again the heterozygote may have a resistance to malaria.

The $\beta$-globin gene is present in $\beta$-thalassaemia but the erythroblasts either fail to produce functional $\beta$-mRNA (severe forms) or they synthesise sub-normal amounts (milder forms). This could reflect faults at one of several stages in protein biosynthesis: transcription of the DNA, processing of the primary transcript RNA or translation. In $\beta$-thalassaemia reticulocytes synthesise $\alpha$-chains, which precipitate in the absence of $\beta$-chains. The cells do not complete their maturation and are destroyed before leaving the bone marrow. The condition only becomes apparent after the first few months of life, during the transition from HbF ($\alpha_2\delta_2$) to HbA ($\alpha_2\beta_2$). Some compensation is obtained through childhood by a continued synthesis of HbF but, if untreated, survival is rarely beyond the second decade. The affected child requires frequent blood transfusions with the consequent risk of iron overload (see below).

The situation in α-thalassemia is slightly different. Because the α-globin locus is duplicated, a normal individual has four α-genes. Deletion of all four genes is fatal and the fetus dies *in utero*. With one functional gene most of the haemoglobin produced has four β-chains and is very unstable; it precipitates in the erythrocytes causing haemolytic anaemia. The presence of two or three α-globin genes does not normally result in anaemia.

Several of the rarer abnormal forms of haemoglobin are unstable and precipitate within the red cell forming granules known as Heinz bodies; such erythrocytes are fragile and lyse more readily. One example is Hb$_{Hammersmith}$ where serine replaces a phenylalanine residue (β42) that is involved in haem–globin contacts. The decreased hydrophobicity permits water to enter the haem pocket leading to oxidation of the iron atom. The haem, being less tightly bound, dissociates and the unstable globin precipitates. There are other similar abnormal haemoglobins, and the severity of the disease in each case correlates with the ease with which haem leaves the pocket.

*Abnormal Enzymes.* There are documented inherited disorders of virtually every red cell enzyme and always the outcome is an increased risk of haemolysis. Only one of the diseases, glucose 6-phosphate dehydrogenase (GPDH) deficiency, is at all common, with an estimated 100 million sufferers in the world. The distribution correlates with the incidence of malaria, and once again a similar protective role has been postulated for GPDH deficiency. The GPDH gene is on the X-chromosome, so that all erythrocytes are affected in males and homozygous females, whereas the heterozygous female has one population of erythrocytes with full enzyme activity and the remainder with virtually no GPDH activity.

In the mature erythrocyte GPDH, the first enzyme of the pentose phosphate pathway, provides the NADPH essential for removing activated $O_2$ or $H_2O_2$ thus preventing protein denaturation and consequent haemolysis (see p. 17). Most affected individuals pass through life unaware of a problem because the residual activity of the abnormal enzyme is sufficient to produce enough NADPH for the red cell to function. The deficiency becomes apparent only when the demand on the pentose phosphate pathway exceeds the capacity of the enzyme. The system may become overloaded during an infection or by compounds that increase the formation of $H_2O_2$. Examples of drugs that provoke haemolytic episodes include aspirin, sulphonamides and anti-malarials such as primaquine.

Over 50 different isoenzyme forms of GPDH are known and these, like haemoglobins, arise from amino acid substitutions. Although most isoenzymes show little change in activity, a few variants either are unstable and denature more readily or are catalytically inefficient. The so-called African type has 10 per cent of the normal activity, but the enzyme binds $NADP^+$ more strongly making catalysis more efficient. This effect compensates in part for the decreased amount of enzyme. A severe anaemia occurs with the Mediterranean and Asian forms. The latter variant has only 1–3 per cent of normal activity and because the enzyme is unstable it denatures as the cells age. During haemolysis, older erythrocytes are destroyed initially because young cells contain more functional enzyme.

Deficiencies of other enzymes, apart from pyruvate kinase, are very rare. Glycolytic enzyme deficiences impair energy production, which affects membrane stability through the declining activity of the ion pumps, as well as altering the 2,3-diphosphoglycerate concentration and hence oxygen delivery. Defects in glycolytic enzymes that act prior to the formation of 2,3-diphosphoglycerate inhibit its synthesis, whereas those acting after raise 2,3-diphosphoglycerate levels by preventing its breakdown. Lastly, a deficiency in methaemoglobin reductase can occur; this leads to methaemoglobin accumulation and subsequent cell lysis.

## Acquired Haemolytic Anaemias

Haemolysis can arise for a number of reasons unconnected with the inheritance of abnormal genes. One of the commonest causes throughout the world is malaria with an estimated 250 million cases each year. Other infections also cause haemolysis; for example, the bacterium *Clostridium welchii* releases a phospholipase (also present in spider and snake venoms) that lyses cell membranes.

The survival of erythrocytes is decreased if antibodies are present that bind to red cell antigens. Mismatched blood transfusion (see p. 166) and haemolytic disease of the newborn, are examples of situations where the immune systems activate complement-mediated lysis or stimulate phagocytosis of erythrocytes. Some people possess 'cold' autoantibodies directed against their own red cell antigens. These IgM class antibodies normally cause few clinical problems because they react at temperatures below 30°C. However, should the body temperature fall, for example through poor circulation in the elderly, the antibodies bind to red cell antigens (often those of the I blood group – see p. 107), activate complement and cause intravascular haemolysis.

### Iron Metabolism and Anaemia

*Introduction*

Iron is an essential element; most of it is present in the body as a haem complex in metalloproteins. These *haem proteins* serve many functions. In some instances the iron atom undergoes a reversible oxidation-reduction (e.g. cytochromes). In other proteins the iron is permanently in the Fe(II) state, but the reversible ligand binding properties are utilised (e.g. haemoglobin and myoglobin). Both properties are used in some enzymes either to activate oxygen or to destroy $H_2O_2$, as in oxidases or peroxidases respectively. The remainder of the body's iron is not in haem groups; these *non-haem iron* compounds include ferritin, transferrin and the iron-sulphur complexes of the electron transport chain.

**Table 5.2: The Distribution of Iron in a 70 kg Man**

|  | mg iron | % of total iron |
|---|---|---|
| Haemoglobin | 2,600 | 56 |
| Myoglobin | 400 | 8.7 |
| Cytochromes | 25 | 0.5 |
| Transferrin | 8 | 0.2 |
| Ferritin: liver | 410 | |
|        spleen | 48 | 34 |
|        marrow | 300 | |
|        other tissues | 800 | |

The iron concentration in solution is maintained at a low level. At physiological pH values Fe(II) would be oxidised to Fe(III) with the formation of ferric hydroxide or oxyhydroxide, which precipitate. This is prevented because Fe(II) ions are chelated in organic complexes, normally proteins. The adult male has 4–5 g of iron, distributed between the various proteins as shown in Table 5.2. Only a very small, but extremely important, proportion is present in iron-containing enzymes. The function of the iron stores, held mainly in liver, spleen and bone marrow, is to create a reservoir that buffers fluctuations in availability and demand, normally in relation to haemoglobin synthesis. Aspects of haem synthesis and breakdown are dealt with in Chapter 4.

## Dietary Iron

Each day the red cells from approximately 50 ml of blood are destroyed, liberating 20-25 mg of iron (see p. 37). Virtually all of this iron is salvaged and returned to the body's iron pool. The main causes of iron loss are haemorrhage and destruction of cells lining the gastrointestinal tract. Because the mechanisms for iron retention are so efficient the daily requirement is very low, and a balanced diet contains an adequate supply. Surprisingly for an element of such importance the only control on the amount within the body is through the regulation of absorption from the gastrointestinal tract. No mechanism apparently exists for regulating its excretion. In consequence, excess iron accumulates with serious results, because iron is very toxic.

The normal dietary requirement for iron is set at 10-15 mg per day. During pregnancy, menstruation or phases of rapid growth the requirement may double, a voluntary donation of blood also imposes an additional burden. Should the loss of iron exceed intake over a period of time, anaemia develops (see below). The availability of dietary iron depends on the presence of other constituents. Iron in haem groups is utilised readily whereas that chelated as insoluble complexes with phosphates or phytates (inositol hexaphosphate — present in many plants particularly whole-wheat flour) is not absorbed. Because only 5-10 per cent of the dietary iron is absorbed the recommended intake is much greater than the body's actual requirement (1 mg per day).

## Iron Absorption, Transport and Storage

In the gut free iron is absorbed by the epithelial cells of the small intestine, probably complexed with small anions such as citrate. The iron in haem groups of haemoproteins in the diet is absorbed differently: the haem groups enter intact and are oxidised in the cell liberating the iron atom. Within the intestinal cell all the iron is oxidised to Fe(III) before being bound to ferritin (see below), which serves as a transient iron store before its release into the blood. Because the life of intestinal epithelial cells is only a few days, the accumulated iron is lost when these cells are shed into the lumen. This may represent a regulatory mechanism that compensates for short-term excesses of dietary iron.

Iron is *transported* in the blood bound to the plasma protein, *transferrin*. This glycoprotein (M.W. 76,000) has a single polypeptide chain, folded so as to create two separate Fe(III)-binding sites. The binding of iron occurs concomitantly with an anion such as bicarbonate. Although each site has a high affinity for iron, in a person in iron balance and having a normal plasma transferrin concentration (3.5 g/l) all potential binding sites are not occupied; typically the transferrin is only 25-40 per cent saturated with iron.

Before incorporation into transferrin, any Fe(II) in the plasma is oxidised to Fe(III). The plasma cuproprotein *caeruloplasmin* possesses ferroxidase activity and may perform this function, in addition to its known role as a copper storage and transport protein (see p. 143).

**Figure 5.1: A Flow Diagram of Iron Metabolism**

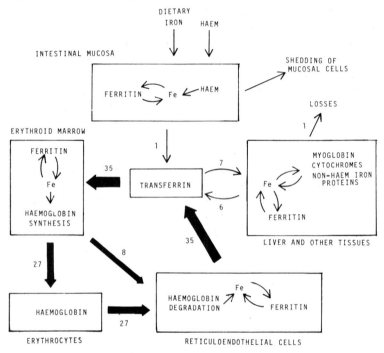

Note: Rates are in mg iron/day.

The iron bound to transferrin is in transit between tissues. Although the amount of iron in plasma is quite small (approx. 8 mg), this pool is drawn upon and replenished continuously. This is apparent from a consideration of the size of the transferrin pool relative to the amounts being transported between tissues (see Figure 5.1). The main route of iron transport is from the liver or spleen to the bone marrow where it is required for erythropoiesis. Transferrin interacts with cell surface receptors on erythroblasts or reticulocytes and transfers the iron atoms to a carrier protein. These receptors are lost when haemoglobin synthesis ceases, so that mature erythrocytes do not scavenge iron from the circulation. Inside erythropoietic cells, the enzyme *ferrochelatase* inserts the iron atoms into molecules of protoporphyrin IX to form haem.

The main *storage* form of iron is the protein *ferritin*. This red-brown coloured cytoplasmic protein is formed from 24 polypeptide subunits, which create a spherical shell of apoferritin surrounding a central crystalline core of ferric hydroxide and phosphate. Each subunit has a molecular weight of 18,500 giving the ferritin molecule an overall molecular weight of 450,000 and a diameter of 12 nm. When fully saturated each ferritin molecule stores approximately 4,000 atoms of iron (about 23 per cent by weight of the molecule). The iron atoms enter or leave the complex via six channels in the protein shell. At some stage between entering the cell and being stored $Fe(II)$ is oxidised; this function is performed by the ferritin, which has ferroxidase properties. Likewise, on leaving ferritin the $Fe(III)$ is reduced. This latter reaction is catalysed by a reductase, which is a flavoprotein that uses NADH as its source of electrons.

In all ferritin-containing cells, another iron storage protein, *haemosiderin*, is also found. Normally, haemosiderin carries 20–40 per cent of the cell's iron stores. It is probably derived from ferritin, but has a greater iron content and forms large insoluble aggregates that are deposited in the cytoplasm. In the event of an overload, the haemosiderin content of liver and other tissues increases, leading to tissue damage. Some of the iron is deposited in the skin causing an ash-grey pigmentation to develop, often one of the first signs of excessive iron deposits.

Iron overload can arise in a number of ways. *Idiopathic* (primary) *haemochromatosis* is an inherited disorder in which affected individuals absorb a larger than normal proportion of iron from the diet. Progressively over a period of years the iron accumulates, particularly in the liver and, if untreated, death occurs, usually through liver failure caused by the deposits. Other people at risk are those who require frequent blood transfusions, for instance those with haemolytic disease, thalassemia or an aplastic anaemia. The extra intake of iron, as an infusion of haemoglobin, bypasses the only control mechanism (i.e. uptake) on iron content. Iron overload created in this way leads to a generalised accumulation of iron throughout the body and death usually results from cardiac failure. Apart from decreasing dietary iron the treatment of iron overload is difficult. The imbalance can be corrected in idiopathic haemochromatosis by removing 500 ml of blood at weekly intervals, but several years may elapse before normal iron levels are restored. Another approach that is suitable for transfusion overload patients is to infuse iron chelators intravenously for several hours each day. One such compound, desferrioxamine, is highly selective for iron and forms a water-soluble iron complex that can be excreted by the kidneys.

## Iron-deficiency Anaemia

If an iron-deficient state is prolonged so that the iron stores become depleted, haemoglobin synthesis is impaired (see below) and anaemia develops. The blood of a person with iron-deficiency anaemia contains smaller erythrocytes (microcytes) than normal, the haematocrit (volume of packed cells) is lower and each red cell has a decreased haemoglobin content. In the developing erythrocyte haem synthesis is geared to haemoglobin production. Consequently, a failure to deliver iron in a sufficient amount retards haem formation in the erythroblasts and leads to an accumulation of protoporphyrin IX that persists in the mature red cells.

The body has a number of compensatory mechanisms which delay the onset of iron-deficiency. In the gastrointestinal tract iron transport is stimulated, increasing the proportion of dietary iron absorbed. The iron-binding capacity of plasma is elevated due to increased production of transferrin, but the percentage saturation is consequently lower. This adaptation aids the salvage of iron by creating more high affinity binding sites.

Iron-deficiency anaemias are most common in infancy and among women of child-bearing age. The fetal iron stores are laid down during the latter stages of gestation. Thus infants born prematurely suffer because they have inadequate stores. Normally, a newborn infant has more red cells per ml of blood than an adult and the haemoglobin concentration inside the erythrocyte is greater. This 'extra' iron is sufficient for development during the first six months of life (milk is a poor source of iron).

In women, menstrual blood loss (on average equivalent to 15–20 mg of iron each month) is the main reason for development of anaemia. If the blood volume lost exceeds 80 ml, a balanced diet would probably not contain sufficient iron to compensate. Pregnancy and childbirth also impose a considerable demand on a woman's iron reserves. During gestation the fetus requires approximately 250 mg of iron, while at delivery a similar amount is lost, largely in the placenta. For these reasons, premenopausal women, particularly during pregnancy, have a greater dietary requirement for iron, which often must be supplemented by iron tablets, typically as $FeSO_4$.

**Anaemias Due to Deficient Haem Biosynthesis**

Apart from iron-deficiency anaemia, other disorders also affect haem biosynthesis. Several inherited diseases are apparently due to deficiencies in the enzymes responsible for protoporphyrin synthesis. These result in the accumulation of iron as cytoplasmic granules in erythroid precursor cells, giving the name, *sideroblastic anaemia*, to the condition.

One of the main effects of *lead poisoning* is inhibition of haem biosynthesis. Lead binds irreversibly to sulphydryl groups, inhibiting enzymes and denaturing proteins containing such groups. Two enzymes that are particularly sensitive to lead are 5-aminolevulinic acid synthetase and ferrochelatase. Their inhibition leads to an anaemia with fewer red cells containing less haemoglobin.

**Folate and Vitamin $B_{12}$ Deficiency**

Normal erythrocyte development depends on the two vitamins, folate and vitamin $B_{12}$ (*cobalamin*). A deficiency in either of these cofactors retards cell division through inhibition of DNA synthesis. This affects all tissues, but is particularly noticeable in those that are rapidly proliferating, such as bone marrow and the gastrointestinal tract. Although cell division and maturation are slower, protein synthesis is largely unaffected so resulting in the formation of fewer cells that are abnormally large. In the bone marrow the large erythrocyte precursor cells are termed *megaloblasts*, giving the name *megaloblastic anaemia*, while in the blood the number of erythrocytes is much lower than normal and macrocytes are formed. However, the haemoglobin concentration per cell is little changed.

*Folate Metabolism*

The active form of folate is tetrahydrofolate ($FH_4$), which is the common name of tetrahydropteroylglutamate. The naturally-occurring forms usually have up to five glutamate residues attached to each other in amide linkages through their $\gamma$-carboxyl groups.

The main dietary sources of folate are fresh green vegatables, although this is supplemented by the synthesis of the vitamin by intestinal microorganisms. The total body store of folate is 10–15 mg of which approximately 2 per cent is lost each day and so must be replaced. A balanced diet contains more folate than the daily requirement, but not a great excess. Consequently, a deficiency can arise quite readily if the diet becomes inadequate or the demand for folate increases through an acceleration of cell division – for example fetal growth during pregnancy or haemolytic anaemia. Therefore, it is not surprising that folate deficiency is the most common form of vitamin deficiency in the world.

Much of the dietary $FH_4$ is oxidised to folate by atmospheric oxygen during cooking and digestion. Also, the polyglutamyl derivative is converted to the monoglutamyl form which is then absorbed in the upper part of the small intestine. Following absorption folate is reduced to $FH_4$ and methylated to methyl-$FH_4$ (see below), which is the form transported in the blood. Once the methyl-$FH_4$ has been taken up by the cells of the body, extra glutamate residues (3–5) are added, which may help to trap the cofactor intracellularly. Epileptic patients on anticonvulsant therapy occasionally develop a megaloblastic anaemia, because the drugs interfere with folate metabolism, probably at the stage of removal of the glutamyl residues in the gastrointestinal tract, thereby impairing folate absorption. Malabsorption is also a problem in several small intestine disorders such as coeliac disease.

There are two aspects to the function of this versatile cofactor. Firstly, it is a carrier of one-carbon units, receiving them from various donor compounds and transferring them to acceptors; secondly, while the carbon unit is attached to $FH_4$ the carbon atom may be oxidised or reduced. Thus a carbon atom may be accepted in one form and transferred in a different redox state. The carbon unit may be attached to either the 5 or 10 positions, or bridged between 5 and 10. The four main forms are 5-methyl $FH_4$, 5,10-methylene $FH_4$, 10-formyl $FH_4$ and 5,10-methylidene $FH_4$; there is a fifth minor derivative, 5-formamino $FH_4$. Together these comprise the 'one carbon $FH_4$ pool'. One of the main sources of carbon atoms for the pool is from serine catabolism which, together with other routes, is summarised in Figure 5.2. On entering the pool all forms, with the exception of methyl-$FH_4$, are interconvertible; the synthesis of methyl-$FH_4$ is essentially irreversible. Two of the most important biosynthetic processes in which folates participate are thymidylate synthesis and purine formation. Consequently, a folate deficiency decreases the size of the pool and impairs the ability to produce the bases for DNA synthesis, which is inhibited.

During the methylation of deoxyuridylate to thymidylate, catalysed by thymidylate synthetase, the cofactor is oxidised to $FH_2$, an inactive form of folate. Reactivation is achieved through reduction of $FH_2$ by a reductase. This enzyme is the target for compounds known as antifolates, such as aminopterin and methotrexate, which have been used in the treatment of leukaemia for many years. They are close structural analogues of folate but bind 10,000–50,000 times more strongly to $FH_2$ reductase than the natural substrates. By inhibiting this enzyme, the recycling of $FH_2$ to $FH_4$ is prevented, causing the folates to accumulate in a non-functional form that is unable to participate in purine or pyrimidine synthesis.

**Figure 5.2: The Pool of One Carbon–Tetrahydrofolate ($FH_4$) Derivatives**

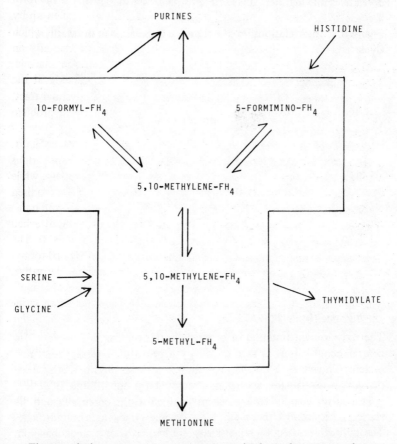

The methyl groups present in many biological compounds are derived from methyl-$FH_4$. The methylation is not a direct transfer, but is via S-adenosylmethionine, an activated form of methionine (see Figure 5.3) in which an adenosyl group derived from ATP is attached to the sulphur atom. After donating the methyl group to an acceptor, the product S-adenosylhomocysteine may be used for the resynthesis of methionine. The reutilisation of homocysteine involves methylation in a reaction which uses methyl-$FH_4$ as a methyl donor, but requires methylcobalamin as a cofactor. It is this connection between folate and cobalamin which is believed to account for the similarity in the types of anaemia that arise from a deficiency of either vitamin. The proposal, known as the 'methyl trap' hypothesis, is that in the absence of cobalamin methyl-$FH_4$ accumulates at the expense of other $FH_4$ derivatives making them unavailable for DNA synthesis.

**Figure 5.3: Interrelationship of Folate and Cobalamin in the Methylation Cycle**

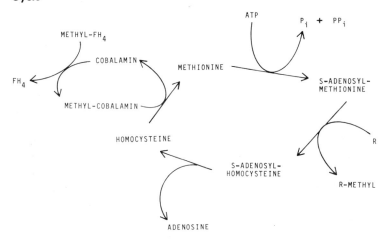

*Vitamin $B_{12}$ (Cobalamin)*

The only known function of cobalt is in the cobalamin complex. The central cobalt atom is surrounded by a substituted tetrapyrrole ring system, known as a corrin ring (similar to a porphyrin ring), which provides four of the six ligands required by cobalt; the fifth is a 5,6-dimethylbenzimidazole covalently linked to the corrin ring and the sixth is occupied by either methyl, deoxyadenosyl, cyanide or hydroxide. The derivatives used therapeutically are hydroxo- or cyanocobalamin. Hydroxocobalamin also can be used to treat cyanide poisoning because cobalamin has a high affinity for cyanide and rapidly exchanges ligands binding cyanide in a non-toxic state. In animal tissues the two forms normally present are deoxyadenosyl- and methylcobalamin, which are also the usual dietary forms. Only two metabolic reactions in mammals depend on cobalamin. The metabolism of propionate requires deoxyadenosylcobalamin, whereas methylcobalamin is an intermediate in the transfer of methyl groups from methyl-$FH_4$ to homocysteine (see above).

The daily requirement for cobalamin is extremely low, a few $\mu$g, and since the body's stores, held mainly in the liver, amount to approximately 3 mg, the effects of a dietary deficiency take several years to become apparent. In developed countries dietary inadequacies are rare; the only group at some risk are vegans because the cobalamin content of green vegetables, fruit and cereals is very low, although even for these people the bacterial contamination of the food may provide sufficient amounts of the vitamin.

Cobalamin in the diet exists complexed with various proteins from which it is liberated by digestion. In the stomach the parietal cells synthesise a glycoprotein, *intrinsic factor*, that specifically binds cobalamin in a high affinity 1:1 complex. This is resistant to proteolysis and thus the cobalamin is protected during its passage along the gastrointestinal tract. Absorption occurs in the terminal ileum where the complex binds to specific receptors on the epithelial cells. Diseases of the terminal ileum, such as coeliac disease, affect absorption and hence give rise to cobalamin deficiency.

*Pernicious anaemia* is a common form (incidence one in 1,000) of megaloblastic anaemia among whites, especially those of Scandanavian origin. In addition to the signs of megaloblastic anaemia, the pernicious form is characterised by atrophy of the gastric mucosa and secretion of a neutral gastric juice lacking intrinsic factor. The absence of the binding protein results in a failure to absorb cobalamin. In most instances, the disease is an autoimmune condition in which the immune response is directed against parietal cells or intrinsic factor (see Table 6.2). In common with other autoimmune diseases, affected individuals often possess the same HLA antigen (see p. 114) and are predisposed towards other autoimmune diseases. Very rarely pernicious anaemia is an inherited disorder affecting intrinsic factor synthesis.

In the blood, two plasma proteins, *transcobalamin I and II*, specifically bind cobalamin. Transcobalamin I has a higher affinity for cobalamin and probably serves a storage role, although transcobalamin II is the protein which initially binds cobalamin and it is the cobalamin-transcobalamin II complex that enters a cell and is catabolised in the lysosomes to release the cofactor. Inside many cells an apparently identical protein to transcobalamin I performs a similar storage function.

## Aplastic Anaemias

The term aplastic anaemia is given to conditions where the red cell count is low because of defective erythropoiesis. In some instances the bone marrow appears normal and the rate of erythrocyte production is essentially unchanged, but the cells are non-viable and are destroyed before leaving the tissue. More usually, the stem cell numbers are substantially reduced, hence the term aplastic, and the defect impairs the differentiation and maturation of all blood cells. The decrease in platelets (*thrombocytopaenia*) affects clot formation, leading to haemorrhage and bruising, and the loss of leucocytes leaves the person more prone to infections. The prognosis for patients with aplastic anaemia is often poor and treatment consists of regular blood transfusions and antibiotic therapy to combat infections. In some severe cases affecting young people, a bone marrow transplant is a possibility if a close tissue match can be made. Since the process of erythrocyte maturation involves so many steps, most of which are not understood, it is not surprising that very little is known about the events causing aplasia. Most efforts have been directed towards identifying and preventing exposure to causative agents. These include:

(1) *Cytotoxic drugs* used for the treatment of leukaemias and other malignant diseases; these act on any rapidly dividing cells. Among such drugs are cyclophosphamide, a DNA alkylating agent; 6-mercaptopurine, an inhibitor of purine synthesis; and methotrexate, an antifolate compound.

(2) A diverse range of other drugs, which can cause aplastic anaemia in a *few* individuals. Among these are the anti-inflammatory agent phenylbutazone, the antibiotic chloramphenicol and the hypoglycaemic agent tolbutamide.

(3) *Chemicals* – notable ones in common use in industry and research are carbon tetrachloride, benzene and DDT.

(4) *Ionising radiations* – radioisotopes such as $^{90}$Sr that are incorporated into the bone marrow present a particular danger to the maturation of blood cells.

## Further Reading

Forget, B.G. (1978) 'Molecular Lesions in Thalassaemia', *Trends in Biochemical Sciences, 3*, 86–90

Friedman, M.J. and Trager, W. (1981) 'The Biochemistry of Resistance to Malaria', *Scientific American, 244*, 112–20

Green, R., Lamon, J. and Curran, D. (1980) 'Clinical Trial of Desferrioxamine Entrapped in Red Cell Ghosts', *Lancet, 8190*, 327–30

Halstead, C.H. (1980) 'Intestinal Absorption and Malabsorption of Folates', *Annual Review of Medicine, 31*, 79–87

Hoffbrand, A.V. and Pettit, J.E. (1980) *Essential Haematology*, Blackwell Scientific Publications, Oxford

—— and Wickremasinghe, R.G. (1982) 'Megaloblastic Anaemia' in A.V. Hoffbrand (ed.), *Recent Advances in Haematology, 3*, Ch. 2, Churchill Livingstone, Edinburgh

Jacobs, A. (1982) 'Disorders of Iron Metabolism' in A.V. Hoffbrand (ed.), *Recent Advances in Haematology, 3*, Ch. 1, Churchill Livingstone, Edinburgh

Maugh II, T.H. (1981) 'A New Understanding of Sickle Cell Emerges', *Science, 211*, 265–7

—— (1981) 'Sickle Cell (II): Many Agents Near Trials', *Science, 211*, 468–70

# 6 WHITE BLOOD CELLS AND THE IMMUNE RESPONSE

## Introduction

Blood contains about $7.5 \times 10^8$ white cells (leucocytes) per dl, although this figure varies widely, even within a healthy individual. Unlike red cells and platelets, white cells are not confined to blood, but are present in the tissues also, where they fulfil many of their important functions; lymph nodes in particular contain large numbers of lymphocytes. There are two broad classes of white cell: the non-granular leucocytes, and the granulocytes, which contain densely staining granules in their cytoplasm (Table 6.1). Polymorphonuclear (PMN) granulocytes are distinguished by the appearance of their nuclei, which have multiple lobes, whereas monocytes have a round nucleus. PMN granulocytes are classified as neutrophils, eosinophils or basophils on the basis of the size, shape, number and staining characteristics of their granules. They may spend only a few hours in blood before entering the tissues.

**Table 6.1: Distribution of White Cell Types**

|  | White cell type |  | % of total white cell population |
|---|---|---|---|
| *Granulocytes* |  |  |  |
|  | Polymorphonuclear granulocytes | Neutrophils | 40–75 |
|  |  | Basophils | Approx. 1 |
|  |  | Eosinophils | 1–6 |
|  | Monocytes |  | 2–10 |
| *Non-granular leucocytes* |  |  |  |
|  | Lymphocytes |  | 20–45 |
|  | Plasma cells |  | 0[a] |

Note: a. Rarely seen in blood, but present in the tissues.

Their overall lifespan is probably a few days and they do not return to the blood. Monocytes, which contain fine granules in their cytoplasm, are present in the blood for a few days only. They migrate to the tissues, where they differentiate into macrophages which survive for months or years without re-entering the blood stream. It is difficult to obtain estimates of the lifespan of lymphocytes because they form a functionally diverse group of cells; some may survive for several years, whereas others are short-lived. In contrast to granulocytes, lymphocytes re-cycle many times between the blood, tissues and lymph.

## Functions of White Cells

White cells are concerned with the defence of the body. The different white cell types do not function independently, but form part of a co-ordinated recognition and defence system that also involves other cells, such as platelets (see p. 116), or the complement system (see p. 99). White cell types, therefore, are not discussed individually, but rather in the context of their biological functions. This defence consists essentially of two types of immune response, the production of antibodies and phagocytosis.

Lymphocytes are the cells responsible for both short- and long-term immunity. They are divided into two broad classes: B lymphocytes, which produce antibodies, and T lymphocytes, which participate in various cellular immune reactions. The phagocytic response is performed mainly by granulocytes which form a 'line of defence' removing unwanted or harmful materials. This is achieved through the recognition, engulfment and killing of invading microorganisms and viruses, the removal of immune complexes and the digestion of dead or moribund cells. The distinction between the different types of granulocyte is often in terms of the location in which the cells perform their functions. PMN granulocytes occur in both blood and tissues whereas mononuclear granulocytes (macrophages) are confined to tissues only. As an example of this tissue specificity, the splenic macrophages ingest red cell fragments and degrade the globin (see Chapter 4).

## Differentiation of White Cells

White cells originate from precursor cells known as pluripotent stem cells, so-named because they also give rise to red cells and platelets (Figure 6.1). In adults the major site of granulocyte formation is the bone marrow, whereas monocytes arise in the spleen and lymphoid tissue; lymphocytes are formed mainly in the thymus and lymphoid tissue. Red cells and platelets arise in the bone marrow. The factors that govern which pathway of differentiation is followed by a particular cell line are poorly understood. However, the importance of certain factors, such as erythropoietin which may control the formation of red cells, is well recognised. In contrast to red and white cells, which are formed by a process of sequential differentiation of cell types, platelets are derived from large precursor cells, known as megakaryocytes, by repeatedly pinching off cytoplasm. Each megakaryocyte may produce up to 2,000 platelets.

**Figure 6.1: The Differentiation of Blood Cell Types**

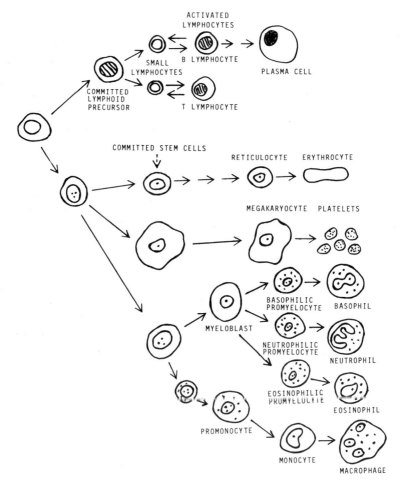

The maturation process of each PMN-granulocyte type is broadly similar, involving a reduction in cell size and a decrease in the number

of most organelles. Mature granulocytes contain relatively few mitochondria, compared with, for example, lymphocytes. Consequently,
granulocytes rely heavily on glycolysis as a source of energy, although
their microbicidal activities do require oxidative metabolism. Their
granules are formed by the endoplasmic reticulum and Golgi apparatus.
The granules of neutrophils and eosinophils share certain properties in
common with lysosomes; for instance, the presence of hydrolytic
enzymes which are involved in the killing and digestion of microorganisms (see p. 85). In contrast, basophil granules lack many of these
enzymes, but are rich in histamine. Monocytes are weakly phagocytic
compared with PMN granulocytes or macrophages. The differentiation
of monocytes into macrophages involves an increase in cell size and
the proliferation of rough endoplasmic reticulum and Golgi apparatus.
Depending on the tissue in which they develop, macrophages may
possess certain characteristic cytological and biochemical features. For
instance, alveolar macrophages contain large numbers of mitochondria
and have an aerobic metabolism unlike other macrophages, such as
peritoneal macrophages and hepatic Kupffer cells, which rely less on
oxidative metabolism. However, in all cases the functions of mononuclear granulocytes are very similar to these of PMN granulocytes.

The committed lymphoid precursor cells (Figure 6.1) differentiate
along two distinct pathways leading ultimately to the formation of T or
B lymphocytes. Pre-T lymphocytes differentiate in the thymus under
the influence of thymic hormones such as thymosin and thymopoietin I,
to give mature T lymphocytes, whereas pre-B lymphocytes develop
in lymphoid tissue associated with the bone marrow. The committed B
or T lymphocytes migrate to lymph nodes, spleen and blood, and do
not undergo further cell division or differentiation until they are
stimulated as part of the immune response. Several rare disorders
are known where B and T cells are missing, although other blood cells
are present. The defect appears to be in the maturation of lymphocytes
from stem cells. Patients with these disorders are very susceptible to
microbial infections.

## Immune Response System

The immune response system contributes to the body's defence against
infection, and is concerned with the recognition of tumour cells. In
responding to a challenge the immune response system is able to
distinguish the body's own cells and components (*self*) from cells and

substances that are foreign (*non-self*), for example, bacteria, fungi, viruses, bacterial toxins and proteins not characteristic of the host. The mechanism whereby cancer cells or those infected with virus are recognised is known as *immunological surveillance*.

An agent that initiates a response is called an *antigen* (derived from the term *anti*body-*gen*erating) — that is a substance capable of stimulating the production of *antibodies* (*anti*-foreign *bodies*). Nowadays the term has a wider usage and is applied to any agent which triggers the immune response system whether or not the synthesis of antibodies is involved.

There are two different types of response, referred to as the *humoral* and *cellular* (or *cell-mediated*) responses. The humoral response, which is directed mainly against toxins, viruses and bacteria, is performed by the B lymphocytes. Within a few days of exposure to antigen, B lymphocytes differentiate into plasma cells, which secrete antibody directed against the particular antigen. The antibody binds the antigen and neutralises it, forming an *immune complex*, which is phagocytosed and digested by cells such as neutrophils or eosinophils. The cellular response involves the T lymphocytes and includes the processes of cell recognition and destruction, as well as immunological surveillance. Although for convenience the functions of these two processes are discussed separately, it must be remembered that the B and T lymphocytes do not act independently, but cooperate with each other in meeting a challenge. For instance, most B lymphocytes exposed to antigen require assistance from certain types of T lymphocytes (T-helper cells) before they will respond by producing antibodies.

The humoral and cellular responses both exhibit the property of *memory*. This arises because some lymphocytes (*memory cells*) are 'programmed' to remember the challenge, and subsequent exposure to the same, or closely related, antigen provokes an enhanced response. Therefore, once a disease has been contracted, protection is afforded against a subsequent infection. This is the basis of *immunity*. The first exposure need be to only very small amounts of antigen; this is sometimes deliberate, when it is known as *immunisation*. The normal means of immunisation is by injection of, for example, heat-killed bacteria, inactivated toxins or live bacteria of a related, but less harmful, strain, although some vaccines (e.g. polio) are absorbed from the gastrointestinal tract and, therefore, can be administered orally. This type of immunity, known as *active immunity*, in which the individual's immune response system has been stimulated, should be distinguished from *passive immunity*. In this process immunity is conferred either by the transplantation of lymphocytes from an immune to a non-immune person, or by injecting antibodies, which may be of human or animal origin. Antibody administration confers immediate immunity on the recipient, but it is short-lived, because the antibodies or lymphocytes are foreign and so are destroyed. A related form of passive immunity is received by the fetus and the new-born baby from its mother. Some antibodies can cross the placenta, while others are absorbed from the milk via the gastrointestinal tract (see p. 99). This is important during the first few months of life, because, although the cellular response operates normally from birth, the humoral response is not fully developed. Antibodies of the IgM class (see below) cannot cross the placenta and their absence from the blood of the new-born leads to impaired phagocytosis as a result of reduced opsonisation (see p. 86). Infants are susceptible, therefore, to bacterial infections.

## Humoral Response

The antibodies, or *immunoglobulins*, are secreted by plasma cells, derived from B lymphocytes that are located mainly in lymph nodes. Immunoglobulins are a family of glycoproteins possessing certain common structural features; they are divided into five broad classes known as IgA, IgD, IgE, IgG and IgM (described in Chapter 7). Throughout life a person is exposed to a wide variety of antigens and consequently produces antibodies with a correspondingly wide range of specificities; of the some $10^{20}$ immunoglobulin molecules in the circulation there are an estimated $10^5$ -$10^8$ different specificities. The antibody molecule possesses at least two identical antigen-binding sites (i.e. it is *multivalent*). Due to its multivalent nature each antibody can combine with more than one antigen, which may lead to the formation of a precipitate (*precipitin reaction*). If the antigens are part of a cell's surface the antibodies usually cause those cells to clump (*agglutination reaction*). Each binding site on the antibody is directed against a small area a few $nm^2$ in size on the surface of a macromolecule or organism. This area is known as an *antigenic determinant*. For most macromolecules or cells the surface will have several antigenic determinants comprising different molecular features. Consequently the injection of a complex antigen leads to the synthesis of a spectrum of antibodies with differing specificities and a range of binding affinities for the antigen. Each immunoglobulin is synthesised by a specific *clone* of B lymphocytes. Thus, for the immunoglobulins of numerous different specificities there are a similar number of clones of cells, each responsible for the elaboration of a particular antibody. A single antigen selects and stimulates the clone that produces antibody with the appropriate specificity. A complex antigen with several antigenic determinants triggers those clones responsible for the synthesis of various antibodies directed against the antigen. This process is known as *clonal selection.* There are believed to be $10^5$-$10^8$ different clones of cells. Since there are $10^8$-$10^{12}$ lymphocytes, this means that some clones are very small whereas others are large.

Small molecules, such as drugs and steroids, in themselves are incapable of eliciting an immune response. However, if the small molecule is linked covalently to a protein, and this conjugate is injected, antibodies are raised against the protein. Some of these antibodies will be directed against that part of the protein surface to which the drug or steroid has been attached. Once formed, these antibodies react with the small molecule, even when it is no longer linked to the protein. These small molecules are termed *haptens.* Although hapten reactions are of little significance *in vivo*, they are exploited for the raising of antibodies for use in radioimmunoassay and related analytical techniques (see Chapter 12).

The response of a clone of cells to its particular antigen proceeds as follows. Within the membrane of the B lymphocyte are approximately $10^5$ receptors with binding sites that recognise antigen. These receptors are apparently antibody-like molecules, which contain an additional hydrophobic sequence; this results in it being embedded in the plasma membrane. The receptor has the same specificity as the antibody secreted by that clone. Binding of antigen to these receptors initiates several events in the B cell. One of the first is a rapid, energy-dependent migration of the receptor-antigen complexes to one end of the cell, a phenomenon known as *capping.* Shortly after this a proportion of these complexes are engulfed, whilst the remainder are shed from the cell surface. The internalised antigen is associated with the nuclear material, but the precise significance of this observation is not understood. One may speculate that it is concerned with an alteration in the control of genetic transcription or mRNA processing leading to a change in the state of cellular differentiation.

The response of the B cell may be influenced by T cells. *T-suppressor cells* prevent the B cell from performing any further response, for example as should occur if a B cell recognises a self antigen. The mechanism of this suppression is not clearly understood, but may involve either the elimination of that clone of cells or the blocking of its surface receptors for antigen thereby preventing it from being stimulated to synthesise antibody. *T-helper cells* may promote the production of antibodies in B cells by performing the initial recognition of certain antigens. The mechanism whereby the helper cells communicate with the B cell may be via the production of a protein that behaves like a specific hormone; or there is a direct membrane-membrane interaction between the two cells. The recognition of antigen and stimulation of B cells also require the assistance of macrophages. The precise way in which the three cell types interact and the nature of the biochemical signals involved are not clear. Some B cells act independently of T cells, for example those involving IgM production in response to high molecular weight polymers with regular repeating structures, as in bacterial polysaccharides.

Regardless of the method of stimulating the B cell it responds by increasing RNA, DNA and protein synthesis, and energy consumption. The number of lymphocytes within the clone increases rapidly, and the cells differentiate to form two types — *plasma cells* or *memory cells*. This process takes about five days and involves some eight cell generations. The plasma cell synthesises and secretes antibody to combat the immediate challenge, whereas the memory cell incorporates antibody into its plasma membrane ready for the next challenge. All of these daughter cells synthesise antibody of the same specificity. Initially, plasma cells generally synthesise antibodies of the IgM class. However, as the immune response develops so the clone may switch to production of other classes, but all of these are directed against the same antigenic determinant. As the cells differentiate their size increases and the endoplasmic reticulum proliferates. At this stage the cell produces essentially two types of mRNA, those coding for the two polypeptide chains of the antibody (see Chapter 7). These are translated by membrane-bound ribosomes and extruded through the endoplasmic reticulum into the cisternal space where the chains associate and stabilising disulphide bridges are formed. During its passage through the membrane, carbohydrate is added to the polypeptide (all immunoglobulins are glycoproteins); this is a two-stage process and is completed in the Golgi apparatus. The immunoglobulin migrates to the cell surface where it is discharged (plasma cells) or remains embedded in the membrane (memory cells). Plasma cells live for only a few days, but in that short time they produce and secrete immunoglobulins at a very high rate, up to 30,000 antibody molecules per cell per second. In contrast, memory cells survive for many years. The increased population of memory cells after the first exposure to a particular antigen explains why on subsequent exposure to the same antigen the response is quicker and the antibody is made in larger amounts. Although memory cells are long-lived, they are not immortal. For this reason immunisation has to be repeated periodically.

In *agammaglobulinaemia*, a rare congenital disease, the individual has a normal cell-mediated response, but does not secrete antibodies. This condition is generally not apparent for the first few weeks of life, due to the maternal transfer of IgG, but as these maternal antibodies disappear the resistance to bacterial infection decreases, mainly as a result of impaired opsonisation (see p. 86). A related, but less severe, condition is *dysgammaglobulinaemia*, in which one or two classes of immunoglobulin are abnormal or absent; a compensatory increase in the synthesis of other classes may occur.

A balanced synthesis of the two polypeptide chains of the antibody is essential. Overproduction of light chains relative to heavy chains leads to secretion of free light chains into the blood. These light chains are small enough to be removed by the kidneys and thus appear in the urine where they are referred to as Bence-Jones proteins (see p. 93). Conditions where such an imbalance of synthesis occurs, or where there is unrestrained synthesis of a particular antibody, are termed *multiple myelomas*. A recent advance in immunology is the use of myeloma cells for the production of *monoclonal antibodies*. In this technique a single clone of B cells (which normally cannot be cultured) is fused with a myeloma cell type to produce a cell line which synthesises in an unrestrained manner the immunoglobulin specified by the B lymphocyte (see Chapter 13).

### Cell-mediated Response

The T lymphocytes circulating in the blood and lymph, which are responsible for the cellular reaction of the immune response system, recognise cells through their surface components. The proteins, glyco-proteins and polysaccharides in the outer surface of a bacterial or protozoal cell usually differ markedly from those of the infected host. For cells of the same or closely related species, surface glycoproteins known as *histocompatibility antigens (HLA)* (see Chapter 8) form an important part of the recognition process. The problem of rejection of a transplanted organ or tissue grafts arises because of these antigens, and the host and donor tissue types must be matched in terms of HLA groupings to avoid rejection. Even so, immunosuppressive drugs (e.g. azathioprine and prednisone) or antilymphocyte globulin are used to prevent proliferation of those T lymphocytes which recognise small residual differences in minor surface antigens. Cell-surface glycoproteins, and particularly glycolipids, may be changed in malignant cells, and it is these altered surface properties that are believed to be so important in determining the invasiveness and patterns of distribution of metastases in cancer. Likewise, viral infection of cells often leads to the insertion of virus-specified (glyco)proteins in the plasma membrane; they are recognised by lymphocytes involved in destroying the infected cells.

Functionally T lymphocytes are a heterogeneous group of cells comprising:

(1) *Cytotoxic* or *killer lymphocytes* ($T_C$ cells). These are responsible for recognising and destroying non-self cells.

(2) *Suppressor lymphocytes* ($T_S$ cells). They regulate the magnitude of the response by B or T lymphocytes in the humoral and cellular responses respectively.

(3) *Helper lymphocytes* ($T_H$ cells). These are concerned with assisting B cells in the humoral response, and co-operate with $T_S$ cells in regulating the cellular response. It is possible that these two functions are performed by distinct sub-populations.

The cellular response to antigen proceeds in the following way. The surface receptors of the T cells that recognise a particular antigen strongly resemble immunoglobulins and are probably closely related to them. In much the same way as described for the humoral response there are clones of T cells possessing different receptors, each clone being specific for an antigenic determinant. A clone of T cells stimulated by its antigen proliferates into effector cells and memory cells; the effector cells comprise $T_H$, $T_S$ and $T_C$ cells. The $T_C$ cells attack and lyse foreign cells by an unknown mechanism. Neither complement nor antibody is involved in the lytic process, but the co-operation of macrophages and $T_H$ cells is important. The macrophages are also involved in the digestion of the lysed cell. Stimulated lymphocytes release *lymphokines*, a heterogeneous group of poorly defined effector substances, whose actions include the ability to stimulate lymphocyte proliferation, increase vascular permeability, attract macrophages and promote their phagocytic potential. Lymphokines therefore enhance and localise the cytotoxic reaction. Interferon also may be considered a lymphokine; it is important in the body's defence against viruses and possibly the spread of tumours (see p. 80).

As with most immune reactions, control is exerted by balancing stimulation against inhibition — performed by $T_H$ and $T_S$ cells respectively. The original concept of the immune response was that the body responded to foreign antigens by the formation of antibodies or cell-mediated immunity, but did not respond to its own constituents. The clones of cells responding to self antigens were eliminated or inactivated during early life. This view must now be modified because it is known that on ageing *autoantibodies* are produced reacting against self components. Usually the presence of autoantibodies is not harmful, but in some instances serious disease can result. The occurrence of such conditions highlights the fact that suppression of the reaction against self antigens must continue throughout life. This suppression is achieved normally by an inhibition of $T_H$ cells: the clones of B and $T_C$ cells acting against self antigen are present. A failure in the control of $T_H$ cells or a stimulation of them by an outside factor (e.g. infection or the presence of a similar cross-reacting antigen) leads to stimulation of the particular B or T cells acting against self, resulting in an *autoimmune disease*. Several such diseases are documented, and in each case the body's defence system attacks its own proteins or cells (see Table 6.2).

**Table 6.2: Identity of the Antigens in Autoimmune Diseases**

| Disease | Antibodies directed against |
| --- | --- |
| Rheumatoid arthritis | IgG ($F_c$ region) |
| Systemic lupus erythematosus | DNA |
| Pernicious anaemia | Intrinsic factor |
| Idiopathic thrombocytopaenic purpurea | Platelets |
| Hashimoto's thyroiditis | Thyroglobulin |
| Autoimmune haemolytic anaemia | Erythrocytes |
| Myasthenia gravis | Acetylcholine receptors in muscle |
| Thyrotoxicosis | Cell surface TSH receptors |

Another aspect of immunological control is the phenomenon of *tolerance*, when the immune response system fails to react to certain antigens. This may be expressed as an inability to reject tumours or grafts, or the absence of antibody production. In these instances it is believed that $T_S$ cells are formed, which prevent those clones directed against the antigens concerned from carrying out their normal function.

The extent to which the body reacts to antigenic stimuli is also controlled by genetic factors, but the molecular basis is poorly understood. These factors include the *immune response (Ir) genes* which are closely linked to the HLA genes on chromosome number 6; this region is known as the *major histocompatability complex (MHC)*. It has been postulated that the association of Ir and HLA genes is one of the reasons why particular HLA groupings are related to a predisposition to certain diseases (see p. 114). This genetic region is especially interesting from the point of view of the integration of various aspects of the immune response system. In addition to the HLA and Ir genes, it carries the information for some complement proteins, for transplantation antigens and for the cell surface glycoproteins, characteristic of B and T cells, which are probably concerned with B cell–T cell co-operation in, for instance, immunological surveillance. The recognition by T cells of cell surfaces depends on co-operation between T cells and macrophages. Thus, when the cells of the immune response system 'scan' the surface of a target cell the $T_C$ cells identify the HLA groupings, whereas the $T_H$ cells distinguish the Ia antigens (the products of Ir genes) on the surface of macrophages. The cellular response by T cells will occur if both facets of the recognition process are satisfied. Macrophage and $T_H$ cell Ia antigens similarly are recognised by B cells involved in the humoral response, and may influence the ease with which B cells can be triggered by certain antigens. This may partly explain why two individuals exposed to the same antigen respond to different extents.

A rare congenital condition, the *Di George syndrome*, has been described in which cell-mediated immunity is impaired due to a lack of T cells, but the humoral response is normal. The disease has been successfully treated in some cases with thymic extracts, which suggests that an underlying fault may be a deficiency of thymosin (required for maturation of T cells). In other instances the disease is of embryological origin since both the parathyroid gland and thymus fail to develop. A similar lack of T cells is seen in *thymic dysplasia* in which the thymus is present, but is not developed properly. Patients with these conditions are very susceptible to intracellular infections. Two acquired disorders of the cellular response are known. The first is caused by autoantibodies which, together with complement, lyse T cells. The second is *Hodgkin's disease* (cancer of lymph node cells) in which patients have a normal number of T cells but they are ineffective, because their surface receptors are blocked by serum factors.

**Viruses and the Immune System**

Viral infections are countered by both the humoral and cellular responses of the immune system, although the contribution from each response depends upon the life history of the virus. If free virus particles are present in the circulation, the humoral response is stimulated to synthesise specific antiviral antibodies. These antibodies may lead to direct lysis of the virus through activation of the classical complement pathway, the blocking of virus interaction with its receptor sites or target cells, or aggregation of the virus particles thereby enhancing their phagocytosis. Antibody production is particularly effective against viruses like polio, which have a long incubation period and stay in the circulation for some time; polio virus enters the body through the gastrointestinal tract and reaches its target cells in the brain via the blood.

Many viruses never appear in the circulation, particularly if, like influenza and cold viruses, they initially infect cells close to their entry site into the body. During infection with viruses such as mumps, measles and rubella, virus-coded (glyco)proteins are synthesised by the infected cell and inserted into its plasma membrane. Cytotoxic T cells will recognise these viral antigens and as part of the cellular immune response kill the cell, as long as it also carries the appropriate host HLA grouping.

The body has another important defence against viruses: known as *interferon.* This is a family of glycoproteins synthesised by the reticulo-endothelial system, in particular leucocytes, in response to virus infection. Interferon is species-specific, but *not* virus-specific; that is, human interferon is specific for human cells infected with any virus.

The biosynthesis of interferon is stimulated in cells infected with virus. These cells release interferon, which interacts with receptors on other cells and subsequently protects them from viral infection. The antiviral action of interferon is indirect, since it does not enter target cells, but instead triggers the synthesis of several host-encoded enzymes which inhibit the growth of viruses by preventing their transcription, translation or assembly; the mechanism depends on the virus in question. The most studied aspect of interferon action is the inhibition of protein synthesis in infected cells. Amongst the enzymes whose synthesis is induced are an endonuclease, which degrades mRNA, and a protein kinase, which inhibits protein synthesis by phosphorylating initiation factors.

In addition to its role as an antiviral agent, interferon has other important effects mediated through co-operation with the immune response system, and interferon is produced during immune reactions even when viruses are not involved. For instance, interferon markedly enhances the toxicity of $T_C$ cells towards their target cells. In some diseases, such as multiple sclerosis and myeloma, which involve immune disturbance, there is evidence that decreased interferon production or $T_C$-cell activity might be contributory factors in their pathogenesis. Interferon has been implicated also in autoimmune disease, since a high percentage of patients with rheumatoid arthritis or systemic lupus erythematosus have elevated serum levels of interferon.

When white cells participate in immune reactions they are stimulated to produce so-called *immune interferon*, which differs from that synthesised by virus-infected cells. For example, B cells produce immune interferon after recognising viral antigens, and macrophages, when attracted by chemotactic factors released by $T_C$ cells, will also synthesise immune interferon. It is possible that through its interaction with the immune response system interferon acts, not only as a specific antiviral agent, but as a general mediator of cellular control processes, including those directed against tumour cells. A better understanding of its function has been hampered until recently by lack of purified interferon in adequate amounts. This is now possible with the aid of techniques such as affinity chromatography, so that further structural and mechanistic studies can be performed. The possibility of large scale interferon production for clinical use has been made more likely by the recent cloning of the human interferon gene in bacteria (see Chapter 13).

## Inflammation and Sensitivity

Inflammation is part of the body's characteristic reaction to wounding or infection. It is a complex response, involving most of the types of white cell. *Acute inflammation* is an immediate reaction characterised by an infiltration of PMN granulocytes, and involves the humoral response. This reaction normally lasts for less than two days but, should the stimulus remain, T cells of the cellular response become involved leading to *chronic inflammation*. The increase in vascular dilatation and permeability are largely responsible for the classic symptoms of inflammation, namely redness, swelling, heat and pain. Inflammation is usually a self-limiting process, but in certain instances a breakdown in its control occurs, which gives rise to *sensitivity reactions* or *hypersensitivity states*. Some people become 'sensitised' to *allergens*, for example, dusts, pollen and particular drugs or foods. Upon subsequent contact with allergen their sensitivity can manifest itself in a variety of ways, ranging from a rash to asthma or anaphylactic shock.

The two main mediators of inflammation are *histamine* and the *kinins*, although prostaglandins and related compounds are also involved. Inflammation is initiated in one of several ways:

(a) as part of the immune response system by the release of histamine from mast cells or basophils as a result of either antigen binding to surface receptors or the engulfment of immune complexes;

(b) by the activation of factor XII in blood coagulation during wounding, which triggers kinin production; or

(c) during phagocytosis by granulocytes, which release kininogenase, another mediator of kinin production.

The presence of immune complexes activates the complement system, which in turn attracts phagocytes to the inflammatory site. Thus more than one factor may be involved in the initiation of inflammation.

The *kinins* are a family of closely related peptides, which are potent vasodilatators; they increase capillary permeability and cause pain. The best known of the kinins is the nonapeptide bradykinin. Lysylbradykinin (kallidin) is a decapeptide consisting of bradykinin with a lysine residue at its N-terminus. The pathways leading to kinin production are shown in Figure 6.2. Kinins are formed by the controlled proteolysis of an inactive precursor plasma protein kininogen, of which there are two types.

**Figure 6.2: Pathways of Kinin Biosynthesis**

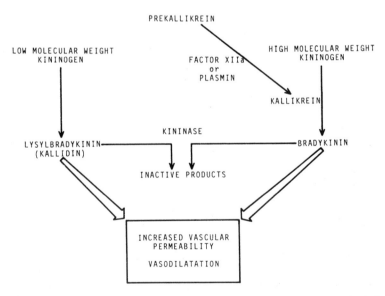

There are two routes of activation: the low molecular weight (LMW) kininogen is activated by kininogenase released from tissue neutrophils, whereas the high molecular weight (HMW) kininogen is activated during initiation of the plasma coagulation system. This generates activated factor XII, which converts inactive prekallikrein to the proteolytic enzyme kallikrein, which in turn liberates bradykinin from the HMW kininogen. Prekallikrein can be activated also by plasmin, the proteolytic enzyme responsible for the dissolution of blood clots. In this way inflammation associated with infection, blood clotting or wound healing leads to the synthesis of a common group of mediators, the kinins.

Histamine potentiates the action of kinins. It is formed from L-histidine in a decarboxylation reaction catalysed by histidine decarboxylase. Mast cells and basophils are particularly rich in this enzyme, and store the histamine in their granules. When basophils bind antigen to their IgE surface receptors, or phagocytose immune complexes, there is a rise in intracellular $Ca^{2+}$ and a lowering of intracellular cAMP levels. These events lead to degranulation and a release of histamine. Blood platelets also contain histamine, which is released during coagulation.

**Figure 6.3: The Biosynthesis of Leukotrienes, Prostaglandins, Thromboxanes and Prostacyclins**

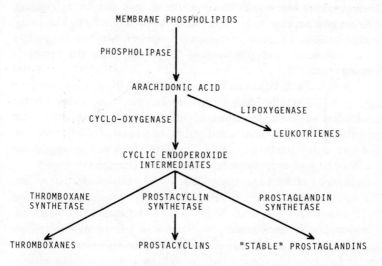

Another class of substances involved in the inflammatory response is the *prostaglandins*, and the related compounds *thromboxanes*, *prostacyclins* and *leukotrienes*. All these are synthesised from polyunsaturated fatty acids (e.g. arachidonic acid) derived from membrane phospholipids by phospholipase action (Figure 6.3). The phospholipids may be part of tissue or blood cell membranes. The polyunsaturated fatty acids are substrates for two enzymes — lipoxygenase and cyclo-oxygenase. Lipoxygenase synthesises a group of hydroxylated polyunsaturated fatty acids called leukotrienes, whereas cyclo-oxygenase produces thromboxanes, prostacyclins or 'stable prostaglandins', depending on the cell or tissue type. The term 'stable' is a relative one, because all these compounds are short-lived, surviving for a few seconds or minutes only.

During the initial stages of inflammation, prostaglandins are made by tissue cells. $PGE_2$ is the major inflammatory 'stable' prostaglandin, and has a powerful vasodilatatory action on tissue capillaries. This causes the redness and oedema so characteristic of inflammation; prostaglandins also sensitise nerve endings and are thus indirectly responsible for the pain of inflammation. White cells, which infiltrate the inflamed area, produce more prostaglandins and maintain the symptoms of inflammation. The white cells are drawn to the inflamed area by various chemoattractants; some of the most powerful are the leukotrienes, released from aggregated platelets in the region. These platelets also produce thromboxane $A_2$, and there is, in addition, a continuous low level production of prostacyclins by the vascular wall endothelium. Thromboxanes are potent vasoconstrictors and platelet aggregators, whereas prostacyclins have equally potent opposite effects. The antagonistic actions of these compounds, together with the instability of thromboxanes and prostacyclins in plasma, limit the extent of inflammation.

Non-steroid anti-inflammatory drugs such as aspirin and indomethacin act by inhibiting the cyclo-oxygenase enzyme, thus preventing the production of these mediators of inflammation. These drugs have potential clinical applications, not only in the treatment of inflammation, but also in the treatment of other conditions in which prostaglandins are involved such as cardiovascular disease (see Chapter 9).

Besides the effects of prostaglandins, thromboxanes and prostacyclins, the duration and intensity of the inflammatory response are controlled by the presence of systems which remove the kinins and histamine. For example, the same neutrophils that release kininogenase to activate the production of kinins also secrete the enzyme kininase that destroys them. Consequently kinins last for only a few seconds in plasma. Histamine likewise is inactivated rapidly by diamine oxidase in plasma. In tissues the mast cells that release histamine also secrete a tetrapeptide which is chemotactic for eosinophils. These produce histaminase, an enzyme that inactivates histamine. Thus the combined activities of enzymes in plasma, white cells and platelets help to mediate the inflammatory response and to prevent it from proceeding unchecked.

Together kinins and histamine are largely responsible for the immediate phase of inflammation, whereas prostaglandins are involved in the delayed phase. The potent degradation systems serve to localise and limit the response. In hypersensitive individuals the response is not contained, and within a few minutes of stimulation there is a massive degranulation of the mast cells as they discharge their histamine. In severe cases this can lead to anaphylatic shock, as a result of capillary vasodilatation, contraction of smooth muscle and constriction of airways in the lungs. The response subsides within a few hours. Allergic conditions such as asthma or hay fever may be controlled therapeutically by antihistamines. One group of antihistamines, for example mepyramine, act by blocking the histamine receptor sites in sensitive tissues. Another group, for example disodium cromoglycate, prevent the release of histamine from basophils or mast cells; they prevent degranulation by stabilising intracellular cyclic AMP levels, probably through inhibition of the enzyme phosphodiesterase that breaks down cyclic AMP.

*Delayed hypersensitivity* is an inflammatory response mediated by the cellular, rather than the humoral, system. Unlike immediate sensitivity, which involves PMN leucocytes, the cells concerned with the delayed reaction are monocytes, macrophages and lymphocytes. The clones of T cells that have memorised a previous exposure to allergen are stimulated to differentiate, proliferate and to release lymphokines which provoke an inflammatory reaction involving macrophages and platelets. Cytotoxic lymphocytes are activated and attack cells bearing allergens. Thus, the course of delayed hypersensitivity is the same as the cell-mediated reaction, except that it is more exaggerated.

## Phagocytosis and Killing of Bacteria

The destruction of bacteria can be conveniently itemised as follows:

  (a) granulocyte migration towards bacteria, known as *chemotaxis*;
  (b) recognition and attachment of bacteria to the granulocyte plasma membrane;
  (c) engulfment of bacteria and formation of the primary phagosome, known as *phagocytosis*;
  (d) killing of bacteria; and
  (e) formation of the secondary phagosome and enzymatic digestion of bacteria.

**Figure 6.4: Diagrammatic Representation of Bacterial Opsonisation and Recognition by a Phagocyte**

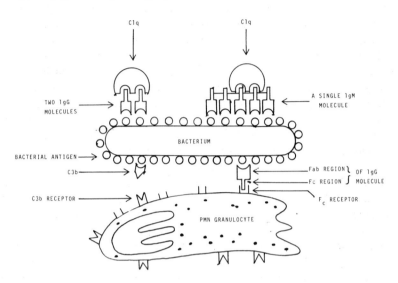

(a) All bacteria produce chemotactic substances which attract phago-cytic white cells. Additionally, certain components of the complement system are chemotactic, for example C3a (see p. 102). Once phagocytosis has begun, neutrophils release compounds, including prostaglandin $E_2$, which draw more phagocytes to the inflammatory site.

Phagocytosis also provides the main control over chemotaxis, because it removes the bacteria and complement factors.

(b) The surfaces of endothelium, blood clots or exposed tissue collagen stimulate phagocytosis. However, the efficient engulfment of bacteria depends upon a process known as *opsonisation* (derived from the Greek word 'opsonin', which means 'to prepare food for'). Antibodies of the IgG and IgM classes bind to surface antigens of bacteria, but only IgG is opsonic (see Figure 6.4). The reason for this is that when IgG is bound its conformation is changed so that sites are exposed which are recognised by phagocytes. The IgG molecule forms a bridge between the bacterium and the plasma membrane of the phagocyte. The IgM molecule also changes conformation when it binds to bacteria, and both bound IgG and IgM activate the classical pathway of complement fixation, leading ultimately to bacterial lysis. IgM is more efficient at binding complement than is IgG. The antibodies present in non-immune serum are mainly of the IgM class. On immunisation, the formation of

IgG antibodies is stimulated, which, therefore, increases the efficiency of opsonisation. In addition, certain components of the complement system are bound to bacteria and promote phagocytosis. The involvement of both antibody and complement in opsonisation and bacterial lysis forms a common link between these two processes.

(c) The factors that mediate and control the formation of the primary phagosome are not well understood. Pseudopodia flow around the bacteria, which eventually become completely enclosed within a membrane-bound vesicle whose inner face was originally the outside face of the plasma membrane. Up to 30 bacteria may be engulfed by a single phagocyte. This process, which can involve the internalisation of up to half the plasma membrane, necessitates an increase in both membrane synthesis and turnover.

(d) Bacteria that have escaped complement-mediated lysis are killed within the primary phagosome. The main mechanism of killing is by activated oxygen. Characteristically phagocytosis is accompanied by a ten- to 20-fold increase in oxygen consumption by the phagocyte. This extra oxygen is used largely to generate $H_2O_2$, not for oxidative phosphorylation. The enzyme NADPH oxidase, within the phagocyte plasma membrane, uses this molecular oxygen to oxidise NADPH on the inner face of that membrane, while the product, superoxide ($O_2^-$), is released from the opposite face of the membrane. The activity of the pentose phosphate pathway increases markedly during phagocytosis in order to provide the necessary NADPH. Superoxide formation is stimulated as soon as the phagocyte makes contact with bacteria. Thus, initially some superoxide is generated outside the phagocyte, but as the plasma membrane invaginates during phagocytosis the activated oxygen is released inside the primary phagosome. Superoxide can take part in various oxidation or reduction reactions, which often lead to the formation of other reactive species, for example singlet oxygen ($^{\cdot}O_2$), hydroxyl radicals ($^{\cdot}OH$) and $H_2O_2$ (Figure 6.5a). All these can kill bacteria directly by reacting with, for instance, proteins, so denaturing them, or by oxidising lipids to lipid peroxides, thereby damaging membranes. The importance of $O_2^-$ and $H_2O_2$ in microbicidal reactions is seen in patients who suffer from the rare X-linked recessive disorder, *chronic granulomatous disease*. Such patients are highly susceptible to bacterial infections, because their phagocytes ingest bacteria but cannot kill them. The basis of this defect is either a deficiency in NADPH oxidase, or an inability to stimulate the enzyme. Whichever the cause, there is no burst of superoxide production during phagocytosis. *Job's syndrome* is a similar disorder inherited as an autosomal recessive.

**Figure 6.5: The Formation and Removal of Activated Oxygen**

(a) Formation

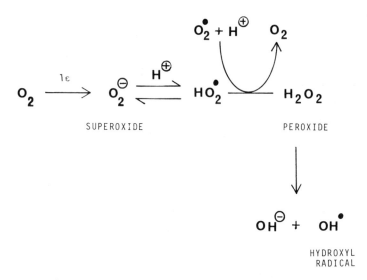

SUPEROXIDE

PEROXIDE

HYDROXYL
RADICAL

(b) Inactivation

The phagocyte must be protected from the harmful effects of $O_2^-$ and $H_2O_2$, which diffuse across the phagosomal membrane into the cytoplasm. Here the $O_2^-$ is converted by the enzyme superoxide dismutase into $H_2O_2$, which then is inactivated either by catalase or glutathione peroxidase (Figure 6.5b).

(e) Once the primary phagosome has been formed, the membranes of granules fuse with the phagosomal membrane to form a secondary phagosome. In this way the battery of granule-hydrolytic enzymes is released into the phagosomal interior where they digest the bacteria. In addition, the granules of neutrophils and eosinophils are particularly rich in the enzyme *myeloperoxidase*, which potentiates the microbicidal effect of $H_2O_2$. In the presence of $H_2O_2$ and a halide, myeloperoxidase forms hypochlorite and other reactive derivatives, which halogenate bacteria, thus killing them. Some of the hydrolytic enzymes (e.g. lysozyme and phospholipases) also contribute to the killing of bacteria. However, despite the impressive array of enzymes present in phagosomes, it is quite common to find that the digestive process is incomplete, even in macrophages which are longer lived than neutrophils. Where the final stages of digestion occur is not known; it may be in subsequent generations of phagocytes, which engulf their dead predecessors.

## Further Reading

Benacerraf, B. (1981) 'Role of MHC Gene Products in Immune Regulation', *Science, 212,* 1229–38
Bodmer, W.F. (1978) 'The HLA System', *British Medical Bulletin, 34*
Cline, M.J (1975) *The White Cell*, Harvard University Press, Cambridge, Massachusetts
Gabig, T.G. and Babior, B.M. (1981) 'The Killing of Pathogens by Phagocytes', *Annual Review of Medicine, 32,* 312–26
Herberman, R.B. and Ortaldo, J.R. (1981) 'Natural Killer Cells: Their Role in Defenses Against Disease', *Science, 214,* 24–30
Kerhl, F.A., Jr and Egan, R.W. (1980) 'Prostaglandins, Arachidonic Acid, and Inflammation', *Science, 210,* 978–84
Murphy, P. (1976) *The Neutrophil*, Plenum Press, New York
Roitt, I.M. (1980) *Essential Immunology*, 4th edn, Blackwell Scientific Publications, Oxford
Rose, N.R. (1981) 'Autoimmune Diseases', *Scientific American, 244,* 70–81
Snyderman, R. and Goetzi, E.J. (1981) 'Molecular and Cellular Mechanisms of Leukocyte Chemotaxis', *Science, 213,* 830–7

# 7 IMMUNOGLOBULINS AND COMPLEMENT

## Introduction

The immune-response system recognises, combines with and destroys antigenic material. The important features and nomenclature of the immune-response system are described in Chapter 6. Readers are advised to consult this latter section first in order to obtain the background information necessary for a full understanding of this chapter, which describes those characteristics of immunoglobulins (antibodies) and the complement system that are responsible for some of these immune effects.

During the humoral response, immune complexes are formed between antigens and immunoglobulins. Binding is between specific regions of the immunoglobulin molecule and certain surface features (antigenic determinants) on the antigen. Phagocytic cells may then bind to such immune complexes via the immunoglobulin components and subsequently ingest and degrade them. Binding of some immunoglobulins to foreign cells results in the activation of complement. This activation has several consequences: (a) generation of a protein complex which lyses the cell; (b) initiation of a local inflammatory response; and (c) production of a protein, which is inserted into the bacterial cell membrane in order to promote destruction by phagocytic cells.

There is a great diversity of antigens and thus of antibody specificity, and any consideration of the molecular features of antibodies must account for the generation of this diversity.

## Immunoglobulin Structure

Immunoglobulins represent some 11 per cent by weight of the plasma proteins. There are five different classes of immunoglobulins, which have the same basic structure consisting of two types of polypeptide chains. These chains differ considerably in molecular weight and are known as the light (L) and heavy (H) chains. The immunoglobulin classes are characterised by the type of H chain they possess. The nature of the heavy chain also determines the biological properties. Within each class there are two different types of light chain possible

**Table 7.1: Characteristics of Immunoglobulin Classes**

| | Immunoglobulin class | | | | |
| --- | --- | --- | --- | --- | --- |
| | IgG | IgA | IgM | IgD | IgE |
| *Physical characteristics* | | | | | |
| Heavy chain | $\gamma$ | $\alpha$ | $\mu$ | $\delta$ | $\epsilon$ |
| Light chain | $\lambda$ or $\kappa$ | $\lambda$ or $\kappa$ | $\lambda$ or $\kappa$ | $\lambda$ or $\kappa$ | $\lambda$ or $\kappa$ |
| Structures | $\gamma_2\lambda_2$ | $(\alpha_2\lambda_2)_{1,2}$ | $(\mu_2\lambda_2)_5$ | $\delta_2\lambda_2$ | $\epsilon_2\lambda_2$ |
| | $\gamma_2\kappa_2$ | $(\alpha_2\kappa_2)_{1,2}$ | $(\mu_2\kappa_2)_5$ | $\delta_2\kappa_2$ | $\epsilon_2\kappa_2$ |
| Molecular weight ($\times 10^{-3}$) | 150 | 150–300 | 900 | 150 | 175 |
| Heavy chain MW ($\times 10^{-3}$) | 53 | 53 | 75 | 53 | 75 |
| % of total plasma Ig | 80 | 12 | 8 | 0.2 | 0.002 |
| *Biological properties* | | | | | |
| Macrophage binding | + | − | + | − | − |
| Secreted externally | − | + | − | − | − |
| Bound to mast cells | − | − | − | − | + |
| Located interstitially | + | − | − | − | − |
| Transferred to fetus | + | − | − | − | − |
| Binds complement | + | − | + | − | − |

($\kappa$ or $\lambda$). The principal characteristics of each class are shown in Table 7.1.

The basic structural unit of all immunoglobulins is two identical light chains and two identical heavy chains. These four chains are linked by both intrachain and interchain disulphide bridges (see Figure 7.1).

**Figure 7.1: Structure of Immunoglobulin G Molecule Showing the Location of the Disulphide Bridges**

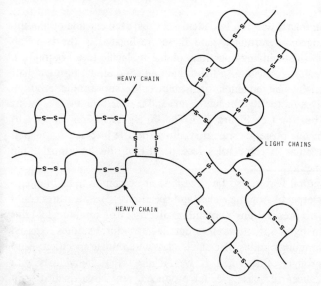

This basic monomeric structural unit is possessed by IgG, IgD, IgE and some IgA molecules. However, IgM and some IgA molecules are multimers, IgM having five monomer units and IgA two. An additional single polypeptide, the J chain, joins the monomer units in these multimers.

**Figure 7.2: Arrangement of Domains and Homology Units in the IgG Molecule**

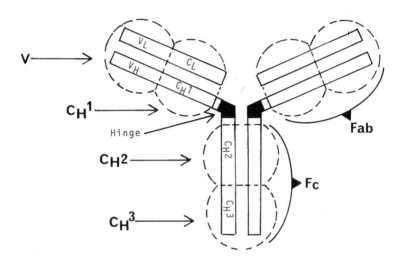

Each monomeric unit has two identical sites, each capable of binding a single antigenic determinant and these are located in the two Fab (*F*ragment *a*ntigen-*b*inding) portions of the molecule (see Figure 7.2). Each Fab region consists of a light chain and the amino-terminal half of a heavy chain. The remainder of the molecule comprises the carboxyl-terminal halves of the two heavy chains and is called the Fc (*F*ragment *c*rystalline) portion (see Figure 7.2). The Fc region has no capacity to bind antigen but possesses sites that allow specific interactions related to the particular effector roles (see p. 98) of the immunoglobulin class concerned.

The structural features so far described are common to all immunoglobulins, irrespective of their class and specificity. Such features could be studied in preparations containing all classes of immunoglobulins possessing many specificities. More detailed structural studies required access to immunoglobulins of a single class and a single specificity, and the breakthrough came when it was realised that individuals with myeloma (cancer of plasma cells) produced very large quantities of immunoglobulin with a single specificity. This represents up to 95 per cent of the total immunoglobulin present and thus was relatively easy to purify. In addition, myeloma patients may eliminate in the urine considerable quantities of the specific light chain of the myeloma protein. These urinary light chains are known as Bence-Jones proteins. Many myeloma-derived immunoglobulin molecules have now been sequenced and a fascinating picture of the different classes and antigen specificities has emerged.

The light chains consist of some 217 amino acid residues. For a given class of light chain ($\lambda$ or $\kappa$) the C-terminal halves have the same or almost identical sequence in all molecules. This is called the constant half of the chain. In contrast, in each molecular species the N-terminal half of the chain possesses a unique sequence, called the variable half. The arrangement of constant and variable regions is illustrated in Figure 7.2. Despite their name, considerable portions of the variable regions are of identical sequence or differ very little from one molecular species to another — that is they have considerable sequence homology. The variations amongst molecules of different specificities are clustered in three particular sequences known as the *hypervariable sequences.*

The IgG heavy chains have approximately 450 amino acid residues. The heavy chain is divisible, like the light chain, into constant and variable portions. In this case the C-terminal three-quarters is constant and the N-terminal is variable. Again the variable regions from molecules of different specificities exhibit considerable sequence homology, but there are four hypervariable sequences within the variable section. About halfway along the heavy chain is the hinge region. This is a short sequence of approximately 20 amino acids, which is rich in proline residues and allows considerable flexibility in this area of the chain.

On the basis of similarities in their amino acid sequence homology the H chains can be divided into four regions, three within the constant portion and one corresponding to the variable portion. These are known as the $C_H 1$, $C_H 2$, $C_H 3$ and $V_H$ sequences (see Figure 7.2). Likewise the L chain can be divided into $C_L$ and $V_L$ sequences. There is homology between all C sequences and separately between the two V sequences. But when the two groups (C and V) are compared there is little homology. These observations on homologous units, together with the arrangement of the intrachain disulphide bonds, suggested that the molecule is organised into *domains* with similar three-dimensional structures. This has been confirmed by limited proteolysis and X-ray diffraction studies. Each domain consists of two homologous units; for example a V domain comprises $V_L$ and $V_H$ units and the $C_H 2$

domain is composed of two $C_H2$ units. The complete domain arrangement for an IgG molecule is shown in Figure 7.3. This is also the arrangement for members of the IgA and IgD classes, whereas IgM and IgE molecules, with their larger heavy chains, have an additional domain consisting of two $C_H4$ units.

**Figure 7.3: The Domains of an IgG Molecule with the Positions of the Hypervariable Regions**

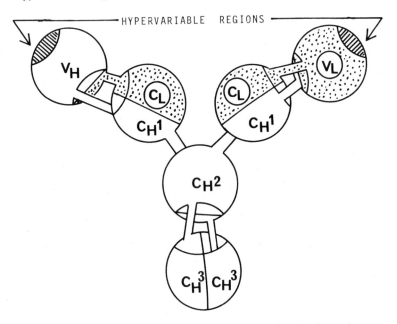

Source: Adapted from R.J. Poljak *et al.* (1972), *Nature, 235*, 137, Figure 3.

An important feature of the structure is that the folding of the domains brings the hypervariable regions of both the light and heavy chain variable sections together within a very small area (a few $nm^2$) at the ends of the V domains (Figure 7.3). An antigen molecule may bind at each of these two sites, the binding specificity depending on molecular complementarity between the site and the surface relief of an antigenic determinant. These sites normally exhibit *multispecificity* — that is they can combine with several different, but structurally-related antigenic determinants. The three-dimensional structure of the binding site for the determinant phosphorylcholine is illustrated in Figure 7.4. In this particular case the interaction between the determinant and its

**Figure 7.4: Diagram of the Binding Site for the Determinant Phosphoryl-choline on an Antibody Molecule**

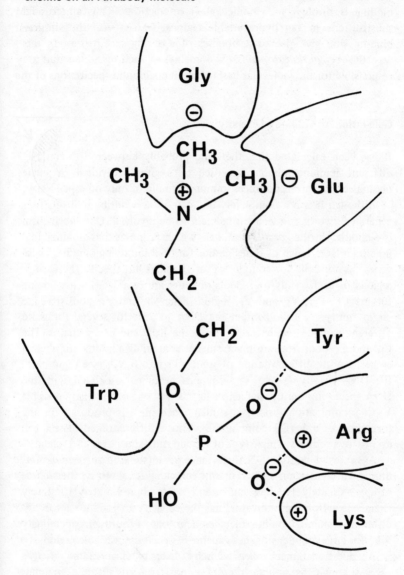

binding site is stabilised by a combination of hydrogen bonds, electrostatic interactions and hydrophobic interactions. In all cases studied the binding is through such non-covalent interactions. Thus amino acid substitutions in the hypervariable regions, which form the antigen-binding site, give rise to antibodies of differing specificities. In this way the *recognition domains* (V domains) of each monomer unit are responsible for the molecular basis of immunoglobulin specificity.

**Generation of Antibody Diversity**

It has been calculated that there are probably between $10^5$ and $10^8$ different immunoglobulins specified by a given individual. A single plasma cell, however, produces immunoglobulin of only one specificity.

Although it is not completely understood how antibody diversity is generated two basic mechanisms seem to be involved. One mechanism is embodied in the *germ-line theory*, which proposes that an individual inherits a large range of genes coding for their immunoglobulins. These genes do not code for complete L or H chains but specify the C or V regions of each. Thus four pools of genes are envisaged, those coding for the $C_L$, $C_H$, $V_L$ and $V_H$ regions. The $V_L$ and $V_H$ pool sizes are large: embryonic cells contain the DNA to code for several thousand different variable sequences for both the light and heavy chains. The $C_H$ and $C_L$ pool sizes are much smaller containing a limited number of copies. During differentiation of stem cells into B cells (see Figure 6.1) the DNA specifying one $V_L$ sequence becomes associated with the DNA specifying the $C_L$ sequence. In this way one of the many possible $V_L C_L$ combinations arises. A similar mechanism is proposed for the formation of a $V_H C_H$ combination. Thus a differentiated plasma cell eventually produces a single type of L chain and a single type of H chain.

As a result of this DNA rearrangement (*somatic recombination*) during differentiation a given plasma cell clone is able to synthesise one of approximately $10^3$ light chains and one of approximately $10^3$ heavy chains. Therefore, an individual has the capacity to produce some $10^6$ different immunoglobulin molecules. Thus one of the origins of diversity is in this large number of genes capable of specifying variable regions.

An additional mechanism whereby diversity is generated is by a process known as *somatic mutation*. As stem cells differentiate into B lymphocytes, mutations occur in the DNA specifying the $V_L$ and $V_H$ regions of the immunoglobulin molecules, thus generating a diversity of light and heavy chain V genes that was not present in the germ line.

The germ line theory outlined above must be modified in the light of recent discoveries about the structure of immunoglobulin genes. The B cells are now known to possess, instead of two pools of genes coding for each chain, three pools of non-contiguous gene segments for each chain. These are referred to as V, J (joining) and C.

**Figure 7.5: DNA Rearrangement and RNA Processing Required for the Expression of the Genetic Information for an IgG Light Chain.** V, J and C represent the variable, joining and constant genes respectively and the two I regions represent the introns. CAP represents the modified 5′ end of the mRNA and $A_n$ the polyadenylic acid 'tail' at the 3′ end.

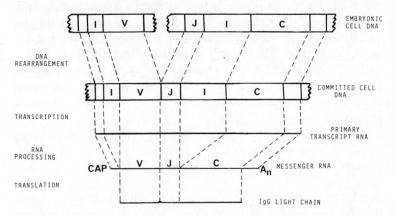

The numbers of different $V_L$ and $V_H$ genes are large — at least several hundred. In contrast there are about ten different $J_L$ and $J_H$ genes, while the number of $C_L$ and $C_H$ genes is smaller. To bring these three coding regions together there is rearrangement of the DNA during differentiation of the B cell (Figure 7.5). For the light chain the first event is the association of one $V_L$ gene with one of the $J_L$ genes. Independent association occurs and thus there is the possibility of one of at least several thousand different $V_L J_L$ combinations.

Because the $V_L$ and $J_L$ sequences both contain hypervariable sequences, light chains with several thousand different binding sites can be generated. Subsequently a second DNA rearrangement occurs and a $C_L$ gene is moved closer to the $V_L J_L$ gene pair. The primary transcript RNA produced in a differentiated cell consists of one $V_L J_L$ sequence combination separated from a $C_L$ sequence by an intervening non-coding sequence. This non-coding sequence is known as an intron and is excised during post-transcriptional RNA processing to leave a messenger RNA with contiguous $V_L$, $J_L$ and $C_L$ sequences. The mRNA is translated into the light chain.

The situation appears to be similar for the $V_H$, $J_H$ and $C_H$ genes of the heavy chain. There is, however, an additional factor to consider. All plasma cells initially produce IgM and subsequently switch to production of immunoglobulin of another class; the antigen specificity does not alter. This switching of classes without change in specificity appears to involve a further DNA rearrangement. To code for the various classes of immunoglobulin requires several $C_H$ genes. Initially a $V_H J_H$ gene pair is positioned close to a $\mu C_H$ gene (which specifies the IgM constant region), so that $\mu H$ chains are produced. During the switch of class the $\mu C_H$ gene is replaced by a $C_H$ gene coding for a different heavy chain class. In this way the class, and hence the biological function of the immunoglobulin, is changed, but since the V regions are unaltered the antibody specificity remains the same.

The precise way in which antibody diversity is generated, and the relative importance of somatic mutation and recombination, is not clearly understood at present, but all these mechanisms appear to be involved.


**Effector Roles of Different Immunoglobulin Classes**

Apart from their function in combining with antigen, antibodies perform several other biological roles. For instance they are involved in phagocytosis and inflammation, and help to protect epithelial surfaces. These *effector roles* are a function of the Fc region of the antibody rather than the Fab region, which is responsible for antibody binding. This Fc region is composed of C domains of the heavy chain only (Figure 7.2). The structure of this region differs from one antibody class to another, and immunoglobulin classes characteristically perform particular effector roles. These include:

(1) *Phagocyte Receptor.* When IgM or IgG molecules combine with antigen the antibody molecule undergoes a conformational change, which exposes sites in the Fc region that bind the immune complex to phagocytic cells (e.g. macrophages). If the antigen is part of a bacterium the process is known as *opsonisation*. The binding to phagocytes stimulates uptake and subsequent destruction of the immune complexes or bacteria.

(2) *Mast Cell Binding Site.* Most IgE molecules are attached to the

surface of mast cells. Only the Fc region of IgE molecules possesses the necessary binding site. When antigen binds to this surface IgE, the mast cell releases heparin and histamine; the histamine causes vasodilatation and smooth muscle contraction. This mechanism is believed to be physiologically significant in promoting the expulsion of parasites from organs surrounded by smooth muscle, for example gastrointestinal tract and uterus. Thus in some gastrointestinal infections IgE may be the cause of sickness and diarrhoea.

(3) *Secretory Component Binding Site.* IgA dimers form a complex with a polypeptide, called secretory component, which facilitates their transport to the epithelial surfaces of the gastrointestinal tract, mammary gland, mouth and skin. There they afford protection by preventing bacterial adhesion to the mucosa. In addition, the presence of secretory component renders the IgA in the gastrointestinal tract more resistant to proteolysis.

(4) *Complement Binding Site.* IgM and some IgG molecules bound to bacteria are capable of activating complement. Activation of the complement system has three consequences:

(a) lysis of the bacterial cell;

(b) promotion of phagocytosis; and

(c) initiation of a local acute inflammatory reaction.

## Complement Fixation and Activation

*Introduction*

As the name suggests, the complement system works in conjunction with the immunoglobulin system. Together they bring about the destruction of foreign cells. Complement consists of some 18 proteins which comprise approximately 2 per cent by weight of the plasma proteins. The proteins of the complement system are organised in two activation pathways, which converge and share a common final lytic sequence (Figure 7.6). In the *classical pathway*, activation is triggered by immunoglobulins coating the surface of foreign cells or bacteria. Complement proteins combine with the immunoglobulin molecules, a process called *complement fixation*, and promote lysis and subsequent phagocytosis of the coated cell. In constrast, the *alternative pathway* can be initiated by the presence of bacteria without the essential involvement of immunoglobulins. The two pathways co-operate in combating an infection: probably, the alternative pathway is triggered initially, with the classical pathway then maintaining the response as antibodies arrive at the site. Although the naming of the two pathways suggests that the classical one plays the major role, this may not be so. The alternative pathway probably evolved as a defence system before there were immunoglobulins, and the classical pathway perhaps arose subsequently.

**Figure 7.6: The Classical and Alternative Pathways of Complement Activation**

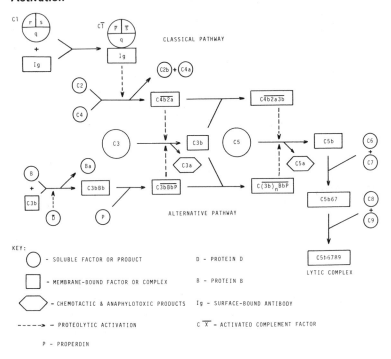

*The Classical Complement Pathway*

The three stages involved in the classical pathway are *recognition*, *activation* and *membrane attack*.

*Recognition.* The first component of complement, Cl, is a protein complex comprising three types of subunit, Clq, Clr and Cls. The Cl complex binds, via its Clq component, to immunoglobulins which have combined with a bacterial cell. The interaction is via specific complement binding sites in the Fc region which are only unmasked after antibody has bound to antigen. Clq must bind to two or more complement binding sites. These sites may be on separate IgG molecules or on different parts of a single pentameric IgM molecule. For this reason IgM is a much more effective initiator of complement fixation than is IgG. For IgG to fix complement at least two molecules must be bound in close proximity. To achieve this hundreds of IgG molecules must bind to an individual cell.

*Activation.* The first event is activation of the proteolytic activity of C1r, as a result of C1q binding to immunoglobulin during the recognition stage. $\overline{\text{C1r}}$ acts specifically on C1s to yield an active proteinase enzyme $\overline{\text{C1s}}$ (Figure 7.6). This is the first of a series of activation steps, which lead to the production of the two active serine proteinases $\overline{\text{C4,2}}$ and $\overline{\text{C4,2,3b}}$, known as C3 convertase and C5 convertase. Each is a complex: C3 convertase consists of activated fragments of C2 and C4, and C5 convertase is produced by the addition of the fragment C3b to the C3 convertase; it is this addition of C3b to C3 convertase that changes its substrate specificity from C3 to C5. Since C3 and C5 are homologous proteins, and a similar bond in each protein is cleaved, the change in specificity is not as great as might appear. These stages, like the analogous system in blood coagulation (see p. 119), allow the initiating event to be amplified and translated into the production of considerable quantities of the protein fragments C3b and C5b. Another important feature of complement activation is that it is confined to the cell membrane, because components C3b and C5b and activated C4 are all membrane-bound.

*Membrane Attack.* Activation of C5 is the last proteolytic step. The final stage in the pathway is the self-assembly of components C6, C7, C8 and C9, with the C5b produced during activation. The formation of this membrane-bound complex apparently creates a hole in the membrane and the presence of just one such complex is sufficient to ensure lysis of a bacterial cell.

### The Alternative Complement Pathway

Activation of the alternative pathway is by a serum glycoprotein, *properdin*. Apparently, properdin is not a proteinase, neither is it activated by proteolysis. It probably undergoes a conformational change upon interaction with IgA class immune complexes, bacterial cell wall components or the presence of bacteria, yeasts or protozoa. Properdin, in combination with complement proteins D, B, and C3b, activates C3 liberating more C3b which maintains the reaction. As with many other components of the complement system, proteins B and D are proteinase precursors, which gain their activity after a limited proteolysis. The uncontrolled consumption of C3 by the positive feedback created by C3b is prevented by the inactivation of excess C3b by a serum inactivator (see below). Having activated C3, the subsequent events are common to those of the classical pathway.

Thus both pathways lead to the formation of a complex which disrupts the integrity of target-cell membranes. There are, however, other products of the complement activation pathways which have important biological functions in phagocytosis and inflammation.

The C3b produced in excess of that required for its role in the activation of C5 binds to cell membranes, where it serves as receptor for phagocytic cells. This ensures the destruction of the cell after lysis. Thus the involvement of C3b in the phenomenon reinforces the phagocyte-binding properties of IgG and IgM.

During complement activation two peptide fragments, C3a and C5a, are released from the N-terminal ends of C3 and C5 respectively. These fragments function as chemotactic agents and attract phagocytic cells to the area. They are also anaphylotoxic, that is they promote the release of histamine from mast cells (see p. 81). This increases vascular permeability facilitating the movement of immunoglobulins, complement proteins and white cells out of the blood stream to the inflammatory site.

## Control of the Complement System

Activation of the complement system is under the control of several different mechanisms. The inherent instability, or low concentrations, of some components limit the response. Thus the activated forms of C2, C3, C4 and protein B have very short life-times, so confining their action to the site of their formation. In addition the plasma contains enzyme inhibitors and inactivators that are specific for some complement enzymes. For example, activated C1r and C1s are inhibited by $\overline{C1}$ inhibitor, whereas C3b is inactivated by a serum proteinase.

Deficiencies in some complement proteins are known to result in increased susceptibility to infections and hypersensitivity reactions. For instance, a deficiency in C3 leads to recurrent infections, while deficiencies in C1r and C2 both give rise to glomerulonephritis and vasculitis.

One of the commonest disorders of the complement system is hereditary angioneurotic oedema (Hane's disease). The molecular basis of the condition is a deficiency of $\overline{C1}$ inhibitor resulting in elevated levels of active C1 in the circulation. The uninhibited C1 activates C2 and C4 liberating a factor that enhances vascular permeability. The net effect of the lack of control over the complement system is recurrent local inflammation. The treatment of the disease is with an androgenic steroid, which increases the synthesis of $\overline{C1}$ inhibitor.

## Further Reading

Amzel, M. and Poljuk, R.J. (1979) 'Three-Dimensional Structure of Immunoglobulins', *Annual Reviews of Biochemistry, 48,* 961–97

Gough, N. (1981) 'The Rearrangements of Immunoglobin Genes', *Trends in Biochemical Sciences, 6,* 203–5

Porter, R.R. and Reid, K.B.M. (1978) 'The Biochemistry of Complement', *Nature, 275,* 699–704

Reid, K.B.M. and Porter, R.R. (1981) 'The Proteolytic Activation Systems of Complement', *Annual Reviews of Biochemistry, 50,* 433–64

Robertson, M. (1980) 'Chopping and Changing in Immunoglobulin Genes', *Nature, 287,* 390–2

Weissman, I.L., Hood, L.E. and Wood, W.B. (1978) *Essential Concepts in Immunology*, The Benjamin/Cummings Publishing Company Inc., California

# 8 BLOOD AND TISSUE ANTIGENS

## Introduction

Exposed on the surfaces of cells are glycoproteins and glycolipids. Regions of these molecules, usually parts of the carbohydrate or polypeptide moieties, are involved in a wide range of cellular phenomena including cell-cell recognition and adhesion, and the specific binding of hormones or small molecules. They are also attachment sites for bacteria and viruses. The structure of surface molecules are determined genetically and are specific for a particular individual. The polypeptide regions of the molecules are specified by structural genes. The oligosaccharide moiety of a glycoprotein or glycolipid is determined by genes that code for the enzymes responsible for assembling the oligosaccharides.

The exposed regions of surface molecules are antigenic and would elicit the formation of antibodies if the cells carrying them were introduced (e.g. during blood transfusion or a transplant operation) into persons not carrying that antigen. It is convenient to divide these antigens into two broad types, the blood group antigens and the tissue (HLA) antigens. The *blood group antigens* are found on the surface of erythrocytes, but many of them are also present on other cell types, and in a few cases the same antigenic determinants occur in secretions. The blood group antigens consist of a limited number of antigenic determinants. This is in marked contrast to the second type of surface markers, the so-called *tissue (HLA) antigens*. An extremely wide range of combinations of these antigens exists so that every person carries on their cell surfaces antigenic determinants that assert their individuality. These antigens are absent from red cells but are in the membranes of other blood cells and most other cell types. The tissue antigens are concerned with phenomena such as immune surveillance and graft rejection (see Chapter 6).

## Blood Group Antigens

Red cells have over 300 different cell surface antigens, which are gathered into 15 *blood group systems*. This classification is based on the closeness of the position and linkage of the genes that code for the antigens. Individuals inherit from each parent a gene, or a linked combination of genes, specifying a blood group, but each blood group system is inherited independently of the others. Thus, each blood group system comprises a series of different, but structurally-related, antigens.

Each blood group system is, in fact, a combination of antigens and antibodies. As well as possessing certain inherited antigens, an individual also produces antibodies directed against many of the antigens he/she does not have. Some of these antibodies are termed 'natural' because they occur without prior immunisation. Other antibodies are 'immune', only appearing in significant quantities after exposure to the appropriate antigen.

There is no systematic method of naming blood groups. Consequently the nomenclature is bewildering and inconsistent. Landsteiner's work at the start of this century led to the discovery of blood groups and the first two alleles were termed A and B. Thereafter other systems were named after monkeys (Rhesus) or the individual who was first recognised as possessing the relevant antibody, for example Mrs Kell (K), Mrs Kidd (Jk) and Mr Lutheran (Lu) (see Table 8.1). Clinically the ABO and Rhesus groups are the most important, although occasionally other groups do cause problems such as transfusion reactions or incompatibility complications during pregnancy. The understanding of blood groups and their serology has been fundamental to the successful development of transfusions.

**Table 8.1: Some Blood Group Systems and Their Commonly Encountered Antigens**

| Blood group system | Major antigens |
| --- | --- |
| ABO | $A_1$, $A_2$, B, H |
| Lewis | $Le^a$, $Le^b$ |
| MNSs | M, N, S, s |
| Rhesus | C, D, E, c, e |
| Kell | K, k, $Kp^a$, $Kp^b$, $Js^a$, $Js^b$ |
| Kidd | $Jk^a$, $Jk^b$ |
| P | $P_1$, $P_k$ |
| Lutheran | $Lu^a$, $Lu^b$ |
| Duffy | $Fy^a$, $Fy^b$ |
| Diego | $Di^a$, $Di^b$ |
| I | I, i |
| Xg | $Xg^a$ |

Surface antigens are usually characterised immunologically and the chemical identities of only a few are known. These include the ABO and Lewis blood groups, and thus this chapter focuses largely on these groups, with a brief review of the biochemical properties of some other blood groups.

## The ABO and Lewis Blood Groups

*Genetic and Immunological Aspects.* In the ABO blood group system individuals are classified into four groups according to whether they possess either the A or B antigens, both antigens or neither — the absence of these antigens is designated O (Table 8.2). These groupings are determined by three allelic genes, A, B and O. The A and B gene products are the A and B antigens respectively. Thus, a person who inherits the A gene from one parent and the B gene from the other synthesises both A and B antigens. Because the O gene is a silent allele (or amorph), someone who is genotypically AO has the phenotype A. These phenotypic relationships are summarised in Table 8.2. Through the use of blood-typing antisera, individuals in group A may be sub-divided into $A_1$ or $A_2$. The antigens formed by these subgroups are identical, the distinction between them being quantitative; an $A_2$ person has a lower density of A antigens than an $A_1$ individual.

**Table 8.2: The ABO Blood Group**

| Blood group (phenotype) | Genotype | Frequency in British population | Predominant antigens | ABO antibodies present in serum |
|---|---|---|---|---|
| A | AA, AO | 42% | A | anti-B |
| B | BB, BO | 8% | B | anti-A |
| AB | AB | 3% | A + B | none |
| O | OO | 47% | H | anti-A + anti-B |

The Lewis blood group comprises two antigens, Le$^a$ and Le$^b$, although these two structures arise through the expression of a single Lewis gene. The majority of individuals have the Lewis gene and so make the Lewis antigens, but the frequency of this gene varies markedly between Caucasian (96 per cent) and black (53 per cent) populations.

The components of the ABO system are strong antigens and the serum of virtually all individuals contains natural antibodies specific for those antigens the person does not possess (Table 8.2). Consequently, if blood from people of the same group is mixed the red cells do not clump, but if, for example, type O serum is added to either A- or B-type cells then agglutination occurs. For this reason ABO compatibility is essential for a successful transfusion. The antibodies are normally of the IgM class and so, in addition to causing agglutination, they also activate the complement system leading to intravascular haemolysis. This reaction starts within a few minutes and is accompanied by fever, fibrinolysis and coagulation. The origin of these natural IgM antibodies, which appear in the first of life, is uncertain. Possibly it is through exposure to ABO antigens, because the chemical structures that form the antigens are very common in nature. The absence of the antibodies towards the individual's own antigens probably results from the particular clone of lymphocytes being suppressed by the continual exposure to the antigen. This phenomenon is known as tolerance (see p. 78). Should a person be exposed to an antigen not normally present on the red cells, for instance by a mismatched transfusion, immunisation occurs and IgG class antibodies are produced.

*Chemical Structures.* The antigens of the ABO blood group system are not confined to the erythrocyte surface but are present on other cell types and in various secretions such as gastric juice, saliva, semen and sweat. The antigenic determinants comprise portions of oligosaccharide units. On the red cell surface some of these form part of glycolipids, but most are in glycoproteins, especially those associated with bands 3 and 4.5 (see p. 9). The density of the antigenic sites on the cell surface is very high, $0.3-1.0 \times 10^6$ antigens per erythrocyte. In the water-soluble forms the oligosaccharides are on the mucus glycoproteins present in the secretion.

The determining factors in the oligosaccharide units are the sugar residues and the manner in which they are linked together. The structures of the five oligosaccharide antigens in the ABO and Lewis blood groups are shown in Figure 8.1. The starting point for assembly of the antigens is the *'precursor oligosaccharide'*, which terminates at its non-reducing end in a galactose residue linked to $N$-acetylglucosamine. In persons of blood group O, fucose is attached to the galactose – this creates the H antigen, the structure synthesised by O individuals. A blood group A person adds $N$-acetylgalactosamine to the galactose of the H antigen, whereas a person of group B completes the chain with a second galactose. People of the AB group synthesise oligosaccharide chains some of which terminate in $N$-acetylgalactosamine and others with galactose. The Le[a] structure is produced by attaching a fucose to the penultimate sugar, $N$-acetylglucosamine, of the 'precursor oligosaccharide'. If both residues of the 'precursor oligosaccharide' are substituted with fucose, the resulting oligosaccharide is the Le[b] antigen.

Recently the 'precursor oligosaccharide' has attracted interest because it is associated with two other antigens, known as I and i, which appear in some types of tumour. These antigenic determinants located in the precursor are normally of minor importance in adults and are detected only in early life. As the ABO antigens develop shortly after birth the I and i antigens decline, probably because the addition of the more peripheral sugars masks the underlying structures. The reappearance of these antigens in later life signifies an alteration in the biosynthesis of the oligosaccharides and can aid in the diagnosis of a tumour.

**Figure 8.1: Biosynthesis of the ABO and Lewis Antigens**

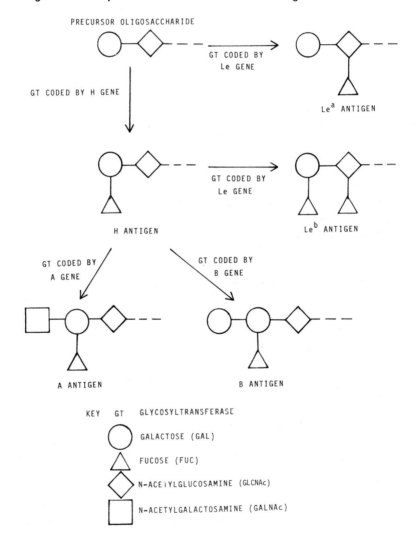

*Gene Activity and Biosynthesis of Antigens.* The oligosaccharides are assembled by *glycosyl transferases*, which add a sugar residue from an activated precursor (e.g. UDP-galactose) onto a particular sugar within the oligosaccharide chain. The specificity of the enzyme determines which sugar is added, its position of attachment and the type of linkage (e.g. $\beta$ 1–3). The glycosyl transferases are coded for by blood group genes: the three independent genetic loci involved in making the A, B, H, Le$^a$ and Le$^b$ antigens are designated ABO, H and Le (Table 8.3).

**Table 8.3: Gene Relationships of the Enzymes Synthesising ABO and Lewis Antigens**

| Locus | Active gene | Reaction catalysed by gene product (glycosyltransferase) | | | Silent allele |
|---|---|---|---|---|---|
| ABO | A | GalNAc attached to Gal | | | O |
| | B | Gal | '' | '' Gal | |
| H | H | Fuc | '' | '' Gal | h |
| Le | Le | Fuc | '' | '' GlcNAc | le |
| Se | Se | — | | | se |

Note: See Figure 8.1 for abbreviations.

The A, B, H and Le genes are termed 'active' and each codes for one glycosyl transferase. In contrast, the silent alleles at these loci are O, h and le, and these do not code for functional enzymes. As far as is known, all individuals are capable of synthesising the 'precursor oligo-saccharide', which is the starting point for assembly of the antigens. The sequential action of the glycosyl transferases converts the precursor into the various antigens (Figure 8.1). Thus, for example, a person of group O who has an active H gene (the vast majority of people in the world have the genotypes HH or Hh) makes the H antigen but does not convert this into either A or B structures. The two Lewis structures are formed by the addition of a fucose to either the precursor (to give Le$^a$) or the H antigen (to give Le$^b$), hence the close relationship of the Le and ABO blood groups.

*Secretors and Non-secretors.* In England, approximately 23 per cent of the population are termed *non-secretors*. In contrast to *secretors*, they do not express their ABO blood group status in their secretions, although they do have the appropriate antigens on their erythrocytes. Whether or not one is a secretor is dictated by the *secretor gene*, which regulates expression of the H gene. In the absence of an active secretor gene the water-soluble mucus glycoproteins do not carry the H antigen and consequently the A and B antigens are absent also (see Figure 8.1). In contrast the H, A and B antigens on red cells are synthesised. Consequently the ABO blood group status of a non-secretor can be established only by testing the blood.

The physiological function of the ABO system is obscure. However, much is known about its biological consequences. The presence of a particular blood group correlates with a predisposition to certain diseases; for example, the incidence of duodenal ulcer is 40 per cent greater in group O individuals, with the frequency in non-secretors being even higher. There is also an association of stomach cancer with people of group A. Another interesting observation is that in cancers of the gastrointestinal tract, cervix or bladder there is a progressive loss of A, B and H antigens from the cell surface, and the disappearance of these antigens correlates with the degree of malignancy.

## The Rhesus Blood Group

The Rhesus blood group is so named because it was observed in the 1940s that antiserum raised against red cells from the *Macacus rhesus* monkey also reacted with human red cells. This antigen, common to both species, was present in 85 per cent of the human population. Unlike the ABO antigens the Rhesus antigens are confined to erythrocytes (approx. 20,000 per cell) and apparently are lipoproteins in the cell membrane. Although this blood group is very important clinically, there is, unfortunately, little biochemical information except that the distinction between the antigens probably resides in the amino acid sequence.

The genetics of this group is more complex than the ABO system. Three alleles, Cc, Dd and Ee, code for the antigens and the linkage between them is so close that the genes are usually inherited together — for example CDe. The antigen corresponding to the D gene is the strongest one and the term Rhesus positive (Rh +) or negative (Rh −) refers to the possession of the D antigen. Rhesus antibodies are not normally present and appear only in Rh− individuals exposed to the Rhesus antigen. The two most common causes of immunisation are transfusion of a Rh− person with Rh+ blood or through a Rh− woman carrying a Rh+ fetus. Because the immune antibodies are of the IgG class the haemolytic reaction takes a different form to that of ABO incompatibility; the IgG antibodies are more efficient at initiating phagocytosis than activating complement therefore, the red cell destruction is predominantly in the liver and spleen rather than intravascularly.

The main clinical problem associated with Rhesus incompatibility is haemolytic disease of the newborn. Complications arise when a Rh— mother carries a Rh+ fetus, the Rhesus D antigen having been inherited from the father. At birth, fetal red cells mix with the maternal circulation and provoke an immune response. Progressively over the next few months Rhesus antibodies appear in the maternal blood. If in a subsequent pregnancy, the Rh— mother carries another Rh+ fetus, the IgG immune antibodies cross the placenta and haemolyse fetal red cells causing anaemia, jaundice, brain damage and even death in severe cases.

The mother may derive some protection if the fetal ABO blood group is also incompatible because the maternal ABO antibodies react with and remove fetal red cells before the Rhesus antigen stimulates the immune system. Another method of protecting the mother is to create passive immunity by injecting anti-D antiserum immediately after delivery. The antibodies, raised in male volunteers, combine with and neutralise the Rhesus antigen. Should haemolytic disease develop in a fetus then exchange transfusion may be required, either *in utero* or after birth. The purpose is twofold, firstly to replace a large proportion of the fetal cells with those devoid of the antigen and secondly to remove the toxic bilirubin from the circulation.

The Kell blood group antigen, K, poses similar clinical problems to the Rhesus D antigen in that antibodies, if present, are immune and cause haemolytic disease. However, the K antigen is much less common and less potent so the problem is rarer.

## Other Blood Groups

Few other groups are important clinically and in most instances the biochemical understanding is meagre. Many of these minor groups are interesting for their anthropological associations. To use two examples: the Di$^a$ antigen (see Table 8.1) is confined to the Mongolian peoples in China and Japan and the South American Indians; the Duffy antigens are present in virtually all white populations but absent from most West Africans.

Biochemically the *MNSs blood group* provides further insights into the nature of antigenic determinants. These antigens, defined by the two alleles MN and Ss, are present on erythrocytes and epithelial tissues. The MN antigens are part of the major erythrocyte glycoprotein, glycophorin A. The specificity resides in the sequence of the five N-terminal amino acid residues of the polypeptide. The M form has the sequence Ser-Ser-Thr-Thr-Gly-, whereas in the N form it is Leu-Ser-Thr-Thr-Glu-. The Ss allele is also expressed in alterations of the glycophorins, but the structural basis is unknown. Although the amino acid sequence is the principal determinant, the glycophorin oligosaccharides also contribute to the reactivity.

As noted above for ABO blood groups, changes in the oligosaccharide structures may occur in tumours. For example, as the MN reactivity decreases another hitherto masked antigen, T, appears. Apart from a potential use in diagnosis, the exposure of this antigen provides a means of attacking such carcinomas. Cytotoxic agents attached to anti-T antibodies can be targeted specifically to the tumour. In addition, these oligosaccharides also serve as receptor sites for influenza virus.

### Histocompatibility and Tissue Antigens

The antigenic determinants that impart a chemical identity to an individual are carried by proteins present on the surface of all cells in the body except erythrocytes and spermatozoa. These antigens are termed *histocompatibility antigens* and their biological functions are all related to aspects of the immune system. Perhaps the most notable consequence of their presence is when incompatibility results in the rejection of a graft or transplant.

The histocompatibility antigens are coded by three main genetic loci – HLA-A, HLA-B and HLA-C – within the *major histocompatibility complex*, which in humans is on chromosome number 6. Within each locus there are a large number of alleles (at present, 20, 42 and eight for HLA-A, B and C respectively); the antigens corresponding to these genes are given a letter/number code signifying the locus/polymorphic specificity, for example A2, B8. Individuals inherit two genes at each locus, one from each parent, and so synthesise six antigens. Because each gene is selected from a large pool, the number of different combinations is very large and the probability of obtaining a perfect tissue antigen match between two individuals is correspondingly remote.

Each HLA gene codes for a protein (molecular weight approx. 43,000) which becomes glycosylated to produce a glycoprotein that is inserted in the membrane. This interacts non-covalently with a smaller protein (molecular weight approx. 12,000), coded by a gene on chromosome number 15; the same protein is also present in serum where it is known as $\beta_2$-*microglobulin*. Thus, an individual synthesises six of these glycoproteins, to each of which is bound an identical $\beta_2$-microglobulin subunit (see Figure 8.2). The HLA antigen is anchored in the membrane by the larger polypeptide, which spans the membrane in a fashion similar to that of glycophorin. In this instance a hydrophobic sequence

**Figure 8.2: Organisation of the Transmembrane Glycoprotein Carrying HLA Antigenic Determinants**

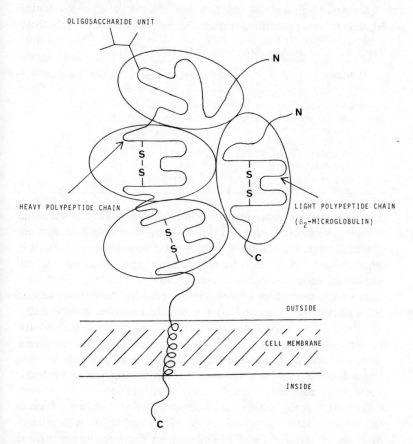

of 24 amino acids connects a relatively short C-terminal intracellular portion with a much larger N-terminal section located outside the cell. The extracellular part is organised in three domains. Two of these domains, like the $\beta_2$-*microglobulin*, are structurally related to an IgG constant domain. These homologies accentuate the close relationship between the various components of the immune-response system. The third domain of the glycoprotein carries the oligosaccharide unit and the antigenic determinants. Antigenic differences between the various polymorphic forms reside in the extracellular part of the polypeptide and consist of multiple amino acid differences rather than single substitutions. It is not known whether the variable regions are clustered in the polypeptide chain, as they are in the hypervariable regions of IgG molecules.

An intriguing aspect of the HLA system is the association of certain HLA genes with a predisposition towards particular diseases (Table 8.4).

**Table 8.4: Disease Associations of Some HLA Antigens**

| Disease | HLA antigen commonly present |
| --- | --- |
| Ankylosing spondylitis | B27 |
| Coeliac disease | Dw3 |
| Juvenile onset diabetes | DRw3, DRw4 |
| Rheumatoid arthritis | DRw3, DRw4 |
| Graves' disease | Dw3 |
| Multiple sclerosis | DRw2 |
| Haemochromatosis | A3 |
| Myasthenia gravis | DRw3 |

These associations do not necessarily mean that an individual will suffer from the disease, only that the chances are greater compared with the population as a whole. Even for a disease such as ankylosing spondylitis, where virtually all sufferers have the B27 antigen, in fact only a small number of people with this antigen are likely to contract the disease. In most instances where a link has been established between tissue type and a disease, the condition has an autoimmune or other immunological association. This is consistent with the clinical observation that auto-immune diseases often occur in families and it has been suggested, therefore, that these particular HLA genes may in some way be involved in regulating the immune response through the suppression of certain T cell clones; suppression is defective in such diseases.

The HLA antigens are directly responsible for regulating certain of the immune activities of the white cells. One way in which this is mediated is through the *HLA-DR antigens* that are characteristic of human B lymphocytes and monocytes. These antigens are coded for by the HLA-D/DR locus of the major histocompatibility complex and participate in the activation of T cells and the initiation of rejection reactions. Like the HLA-A, B and C antigens, the HLA-DR antigens are transmembrane proteins, but their subunit structure is different: they are formed by the non-covalent association of two dissimilar glycoproteins, both embedded in the membrane.

## Further Reading

Dodd, B.E. and Lincoln, P.J. (1975) *Blood Group Topics*, Edward Arnold, London
Ploegh, H.L., Orr, H.T. and Strominger, J.L. (1981) 'Major Histocompatibility Antigens', *Cell, 24*, 287–99
Race, R.R. and Sanger, R. (1975) *Blood Groups in Man*, 6th edition, Blackwell Scientific Publications, Oxford

# 9 HAEMOSTASIS

The circulation of blood serves three vital functions — transport, communication and defence. The interdependence and high metabolic activity of tissues requires that this circulation be rapid and in the closed circulatory system high velocity is achieved as a result of high pressure. Such a high pressure system is especially vulnerable to leakage and it is essential that bleeding due to injured blood vessels is arrested rapidly. The process of haemostasis, which is the spontaneous arrest of blood loss from ruptured vessels, ensures that this occurs. Haemostasis is followed by wound healing.

There is considerable evidence from studies *in vitro* that there exist many potential interactions between the haemostatic, fibrinolytic, complement and inflammatory systems, which are discussed later in this chapter (see p. 123).

## Factors Involved in Haemostasis

Haemostasis involves interactions between the *damaged blood vessel wall*, *platelets* and *circulating blood coagulation factors*. The interactions result in:

    (a) constriction of the blood vessel wall; and
    (b) formation of a haemostatic plug which prevents further blood loss.

There is subsequent slow dispersal of the plug as part of the tissue repair process.

### Damaged Blood-vessel Wall

Blood vessels are lined by a monolayer of endothelial cells. This lining may be supported by layers containing collagen, elastin and smooth-muscle cells. Damage to a blood-vessel wall releases the contents of endothelial cells and also exposes the elastin and collagen fibres of the external layers.

### Platelets

Platelets are small blood cells comprising 0.5 per cent of the blood volume and are synthesised in the bone marrow from megakaryocytes (see Chapter 6). These cells contain large numbers of membrane-enclosed granules and dense bodies, but are not nucleated. Platelets have a high affinity for both collagen and elastin fibres and adhere to them at the site of vessel-wall damage. *Platelet adhesion* is followed by *platelet aggregation* as more cells come together at the same site forming a platelet plug. Another agent which promotes platelet aggregation is thrombin, generated as a result of activation of the coagulation system. The platelet plug contracts and the contents of large numbers of the intracellular membrane-enclosed dense bodies are released within the plug and into the surrounding blood. This is the *platelet release reaction*. Among the contents released are ADP, which promotes platelet aggregation, serotonin (5-hydroxytryptamine) a vasoconstrictor, and calcium ions. Oxidative metabolites of arachidonic acid, including the prostaglandins G2 and H2, and thromboxane $A_2$ (see Figure 6.3), are involved in this second phase of platelet aggregation. They are potent platelet aggregators acting synergistically with ADP. The platelet-release reaction also results in the exposure of platelet phospholipid and factor V, which have important roles in localising subsequent coagulation events.

In small blood vessels platelet adhesion, aggregation and the release reaction play crucial roles in preventing blood loss. The platelet plug seals the damaged wall and serotonin promotes endothelial adhesion and closure of the blood vessel lumen at this site. In larger vessels platelet-plug formation and vasoconstriction are followed by coagulation. The end result of this process is the creation of a haemostatic plug consisting of platelets and fibrin within which other blood cells are trapped.

## Blood Coagulation

Fibrin, the end-product of blood coagulation, is a three-dimensional fibrous network of covalently-linked protein molecules. The mesh of this network is sufficiently small to trap erythrocytes. The events preceding fibrin formation are complex and proceed by one or both of the two pathways illustrated in Figure 9.1.

The sequence initiated by collagen is called the *intrinsic pathway* and involves components normally present in circulation. The *extrinsic pathway* is so called because it involves tissue factors as well as blood components. The two pathways share a common terminal sequence from factor X to the formation of fibrin.

Most of the steps illustrated in Figure 9.1 involve the generation of proteolytic enzymes from inactive precursor proteins (procoagulants).

## Figure 9.1: The Blood Coagulation System

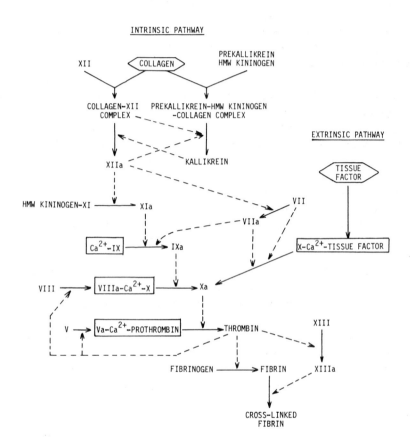

KEY:  HMW KININOGEN – HIGH MOLECULAR WEIGHT KININOGEN

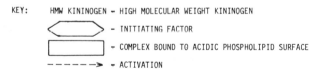

For example, factor Xa is the proteinase generated from an inactive protein, factor X. The proteolytic enzymes generated are of the serine proteinase type; such enzymes have a particularly reactive serine residue present in their active sites. In certain steps, for example the activation of factors XIII, V and VIII and fibrin formation, there is proteolytic cleavage without the generation of a proteolytic enzyme. In these cases there are structural alterations which change the binding characteristics of the molecules or, in the case of factor XIII, result in the activation to a different type of enzyme. All of the factors shown (Figure 9.1) are plasma proteins apart from the initiating factors (collagen, elastin and tissue factor) platelet phospholipid and calcium ions.

The complexity of the system required to generate fibrin merits special comment. The reason for the complexity appears to be that this system operates as a biochemical amplifier. A few molecules of factor XIIa may be produced at the site of damage. These catalyse the production of tens of molecules of factor XIa which in turn catalyse the production of hundreds of molecules of factor IXa. The operation of the subsequent stages amplifies the initial stimulus even more and the end result is the rapid formation of a fibrin network from much of the fibrinogen available at the site of damage. Plasma levels of the various precursor proteins depend upon their positions in the activation pathways. For example, 1 ml plasma contains 3 mg fibrinogen, 100 $\mu$g prothrombin, 10 $\mu$g factor X and <0.1 $\mu$g factor IX. Another property of a multi-stage system is its ability to translate a stimulus into a response which is quite different in character to the initiating event. In this case exposure of collagen and/or release of tissue factor is translated into the formation of a fibrin polymer.

Another complex feature of the coagulation system is the presence of feedback loops and also interconnections between intrinsic and extrinsic systems (Figure 9.1). For example, thrombin and factor Xa serve to activate factors VIII and V; factors XIIa and VIIa are involved at early stages in both the intrinsic and extrinsic pathways.

One feature of the system which serves to localise the generation of fibrin and also to control the degree of amplification is the involvement of platelet phospholipid at two stages: those which generate factor Xa and thrombin. There is some evidence to suggest that when platelets aggregate, those phospholipids that are characteristic of the inner surface of the platelet membrane are exposed. Both factor X and prothrombin bind through $Ca^{2+}$ bridges to this phospholipid (Figure 9.2), thus localising them at the damaged site. The binding sites on factor X and prothrombin consist of several $\gamma$-carboxyglutamic acid residues.

**Figure 9.2: Conversion of Prothrombin to Thrombin.** Factor Xa and prothrombin are bound by calcium ions ($Ca^{2+}$) to the acidic phospholipid-containing surface (▭) through γ-carboxyglutamate-rich regions (◯) near their N-termini. The proteolysis of prothrombin at ➤ by factor Xa is accelerated by the presence of factor Va which is bound to both prothrombin and factor Xa. N and C indicate the amino- and carboxy-termini, respectively, of the polypeptide chains of factor Xa and prothrombin.

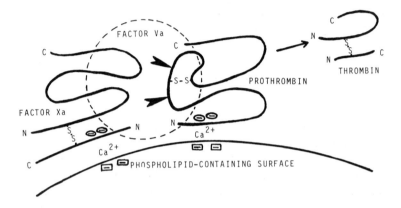

These are introduced, subsequent to protein biosynthesis, by post-translational modification of specific glutamate residues of the polypeptide chains by carboxylation. This is carried out by a vitamin K-dependent enzyme system within the hepatic endoplasmic reticulum. This participation of phospholipid prevents the clotting process spreading too far from the site of damage.

An examination of the step by which thrombin is generated serves to illustrate the complexity of the molecular events which occur in just one stage of the coagulation sequence (Figure 9.2). Factor Xa is bound to the phospholipid-containing surface via $Ca^{2+}$ bridges from γ-carboxyglutamic acid residues localised in the N-terminal region of the molecule. Prothrombin becomes bound in a similar manner along with factor Va, so establishing a factor Xa–prothrombin–factor Va complex on the cell surface. This binding event gives rise to high local concentrations of these components, up to $10^4$ times greater than in plasma. Within this complex, factor Xa hydrolyses two peptide bonds in prothrombin, whereas factor Va behaves as an activator of the process enhancing the rate of formation of thrombin by approximately 350 times. The hydrolysis of the first peptide bond midway along the prothrombin chain yields two polypeptide fragments. The N-terminal half is the $\gamma$-carboxyglutamate-rich region. The C-terminal half of the chain, termed *prethrombin*, is subjected to the second proteolytic step which produces two separate polypeptide chains, A and B, which remain bound together by disulphide bonds. This is thrombin and it is free to leave the surface upon which it is generated. This type of activation, involving limited proteolytic cleavage, is typical of those which occur at other stages in the coagulation sequence, although in other cases the activated molecules may remain bound to the surface, such as in the case of factor Xa.

The final event in coagulation is the conversion of fibrinogen to fibrin. Fibrinogen is a plasma protein (molecular weight 340,000), consisting of three pairs of polypeptide chains, known as $\alpha(A)$, $\beta(B)$ and $\gamma$, which are covalently linked by disulphide bridges ($\{\alpha(A)\beta(B)\gamma\}_2$). The most widely-accepted model for the molecule is an elongated trinodular structure. The proteolytic enzyme thrombin hydrolyses an arginyl–glycine peptide bond close to the N-terminus of each of the $\alpha(A)$ and $\beta(B)$ chains, releasing two pairs of peptides (fibrinopeptides A and B), each of molecular weight approximately 2,000. Although thrombin cleaves just four of the some three thousand peptide bonds in fibrinogen and so has a very limited action on the molecule, the consequences of fibrinopeptide release are, nevertheless, dramatic. The remaining molecule, $(\alpha\beta\gamma)_2$, from which the fibrinopeptides have been cleaved, is called fibrin. Its solubility is very low because fibrin molecules spontaneously aggregate through specific non-covalent interactions to form a highly ordered, fibrous polymer. Not only do fibrin molecules associate to form thick fibres but, more importantly, these fibres branch and create a three-dimensional network with a mesh size small enough to entrap erythrocytes. This network takes the form of a gel which is also capable of retaining for some time the fluid portion of the blood.

Fibrin produced *in vitro* by the action of purified thrombin on pure fibrinogen is soluble in solutions of compounds, such as urea, that disrupt hydrogen bonds. In contrast, a clot produced *in vivo* is not solubilised by urea, because the individual molecules are covalently linked. The enzyme responsible for this, factor XIIIa, is generated from its inactive zymogen by the action of thrombin. The action of factor XIIIa is unlike that of any of the other coagulation factors. It is a transamidase capable of forming isopeptide bonds between specific glutamine and lysine residues in adjacent fibrin molecules. The reaction is illustrated in Figure 9.3. The residues involved are brought together by the specific non-covalent interactions of fibrin molecules. Two such cross-links are made between γ chains in adjacent fibrin molecules near to their C-termini. Further intermolecular cross-links may be made amongst α chains in several fibrin molecules. The result is a covalently-bonded macromolecular network. This network, derived from fibrinogen, which comprises only 0.2 per cent by weight of the blood, is capable, nonetheless, of forming a stable gel which, after cross-linking, has considerable strength.

**Figure 9.3: The Formation of Cross-links Between Adjacent Fibrin Molecules**

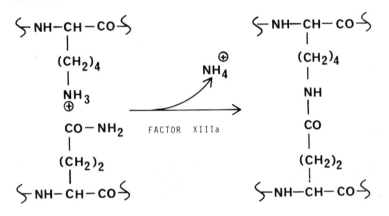

### Control of Blood Coagulation

*Platelet Phospholipid-Coagulation Factor Interactions*
One of the important aspects of control is the involvement of platelet phospholipid in two steps in the sequence to prevent escalation of the coagulation process to sites remote from the injury.

*Proteinase Inhibitors*

Since many of the activated blood coagulation factors are proteolytic enzymes the proteinase inhibitor system of the plasma is of vital importance in the control of blood clotting. The inhibitor system comprises at least six proteins which can combine specifically with certain proteinases to form inactive complexes. Collectively these inhibitors represent a substantial proportion (10 per cent by weight) of the total proteins present in plasma. Probably the most important proteinase inhibitor associated with blood coagulation is antithrombin

III, a molecule capable of combining with, and inactivating, not only thrombin but also factors Xa, IXa, XIa and XIIa. Heparin-like compounds, which are present on the surface of endothelial cells, and heparin, which can be released from mast cells, accelerate considerably the rate of reaction of antithrombin III with these factors (see Figure 9.4). Thus, heparin may be a natural anticoagulant, although there is some dispute over whether it ever occurs in plasma in phsyiologically-significant quantities.

**Figure 9.4: Inactivation of Thrombin by Antithrombin III**

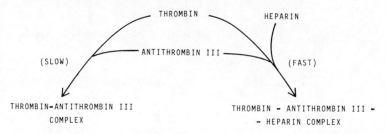

### Fibrinolysis

The fibrin clot is not a permanent structure but is slowly degraded by the action of the proteinase, plasmin, to form a mixture of small and large, soluble peptide fragments which are removed from the circulation by the liver. Plasmin is generated from its inactive precursor, the plasma protein plasminogen, by the action of activators, the nature and number of which are ill-defined. There is preferential generation of plasmin within the clot as both plasminogen and plasminogen activator bind to fibrin. Plasmin generated in this way remains largely bound to fibrin, so confining its action to the blood clot. However, small amounts of plasmin may leach from the clot, or be formed from plasminogen in the general circulation. Since plasmin can degrade both fibrinogen and fibrin, it is important that plasmin in the general circulation be inhibited. This is achieved by the potent inhibitor, antiplasmin, present in plasma.

### Interrelationships Between the Haemostatic, Fibrinolytic, Inflammatory and Complement Systems

These four systems have many features in common: each involves controlled activation of proenzymes to yield serine proteinases; many of the stages are localised by the involvement of membrane-bound cofactors; and additional control is effected by circulating proteinase inhibitors.

**Figure 9.5: Interactions Among the Coagulation, Fibrinolytic, Inflammatory and Complement Systems**

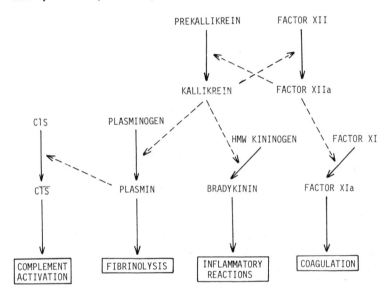

The initial phases in the activation of all these systems appear to be interrelated, as shown in Figure 9.5. There are common components and inhibitors in the early stages of these systems. Thus, deficiencies in plasma prekallikrein, HMW kininogen or factor XII lead to impaired contact-induced coagulation, fibrinolysis, kinin generation and complement fixation *in vitro*.

Factor XII and HMW kininogen both have a high affinity for a variety of surfaces, including exposed collagen. Prekallikrein and factor XI are normally associated with HMW kininogen and, therefore, also become bound. Surface-bound factor XII and prekallikrein may have low proteolytic activity, but sufficient to lead to reciprocal activation of each other yielding factor XIIa and kallikrein. Subsequently, factor XIIa initiates the rest of the coagulation system. Simultaneously it may convert plasminogen to plasmin and initiate the inflammatory reaction by activating prekallikrein. This generation of kallikrein and plasmin may indirectly activate complement fixation. Kallikrein initiates the inflammatory reaction by releasing bradykinin from HMW kininogen (Figure 9.5), but also may be an activator of plasminogen and is capable of initiating coagulation by generation of factor XIIa.

The physiological significance of these connections between the four systems is not clear. Patients with deficiencies in factor XII, prekallikrein or HMW kininogen, all of which are implicated in surface initiation phenomena, do not have any haemorrhagic or thrombotic tendencies. Prekallikrein-deficient patients do not have impaired fibrinolytic or inflammatory mechanisms. Thus, while the potential for interaction between the systems is considerable, their significance *in vivo* has yet to be established.

## Disorders or Haemostasis

The inability to synthesise platelets in sufficient numbers, defects in platelet function or lack of functional coagulation factors can all give rise to uncontrolled bleeding.

### Platelet Disorders

The more common platelet disorders are acquired. For example, aspirin interferes with platelet function (see p. 127). Within two hours of ingesting 1 g of aspirin the bleeding time is almost doubled and this effect may last for several days.

Thrombocytopaenia, a reduction in platelet levels, may be caused by bone marrow diseases such as leukaemia.

### Coagulation Disorders

All genetic coagulation disorders are uncommon, but their consequences are profound. Haemophilia A, in which factor VIII is lacking, is the most common coagulation defect and severe cases occur with a frequency of one in 25,000 of the UK population. Prolonged and recurrent bleeding occurs after minor trauma and is particularly serious if it occurs in the joints and leg muscles. Treatment involves the regular intravenous administration of plasma fractions rich in factor VIII (see Chapter 12). An additional haemostatic agent may be necessary following trauma in haemophiliacs — for example 6-aminohexanoic acid after dental extractions. This compound inhibits plasminogen activation because it is an analogue of the lysine residues in fibrin to which plasminogen binds; thus, 6-aminohexanoic acid slows plasminogen activation and prevents the premature dissolution of fibrin within haemostatic plugs.

Christmas Disease, which is a lack of factor IX, occurs with about 20 per cent of the frequency of haemophilia A. All other genetic coagulation disorders are rarer.

Acquired coagulation disorders can occur as a result of loss of liver function, for example in cirrhosis, since several of the coagulation factors, factors II, VII, IX, X and fibrinogen, are synthesised by the liver.

## Thrombosis

Thrombosis is the formation of a thrombus in the lumen of a blood vessel. It is the result of an imbalance between the process of haemostasis and its inhibition.

### Formation of a Thrombus

Thrombosis may be initiated by changes in the walls of blood vessels and/or, especially in the venous circulation, by cessation or even slowing of blood flow. Changes in arterial walls take the form of atherosclerotic plaques, which are protrusions into the lumen of the blood vessels. These plaques consist mainly of smooth-muscle cells, collagen and lipids.

Venous thrombosis is a particularly common result of physical inactivity and two states where this may occur are, for example, in patients undergoing surgery or in the bed-ridden. Surgery has another effect in that it tends to induce a hypercoaguable state due to platelets with enhanced adhesive properties, and increased amounts of procoagulants but a decreased level of plasminogen. The extent to which a thrombus develops depends upon the velocity of local blood flow, which in turn depends upon the duration and degree of the patient's inactivity.

Arterial thrombosis follows the normal course of haemostasis. Platelets adhere to the arterial wall, forming an aggregate containing some fibrin on which a fibrin tail forms. This tail often contains other blood cells. In contrast venous thrombi contain fewer platelets and tend to resemble normal blood clots.

Thrombosis is a very common cause of death in developed countries, for example, vascular diseases account for some 45 per cent of the deaths in the United States. It is believed that thrombosis is largely preventable by suitable alterations in life-style and diet. High risk factors include a diet which leads to high plasma lipid levels, cigarette smoking, high blood pressure and lack of physical activity.

### Antithrombotic Therapy

Antithrombotic therapy can take three forms. These are: the prevention of platelet aggregation; anticoagulant therapy; and thrombolytic therapy.

*Prevention of Platelet Aggregation.* Agents which modify platelet properties may prove to be the most effective in preventing arterial thrombosis. Four drugs have been found to be effective as anti-platelet agents.

These are aspirin, sulphinpyrazone, dipyridamole and dextran. Aspirin and sulphinpyrazone act on platelet cyclo-oxygenase, the enzyme which catalyses synthesis of prostaglandin G2 from arachidonic acid (Figure 6.3). Both compounds inhibit this enzyme, aspirin by acetylating the active site. Prostaglandin G2 and thromboxane A2 are potent inducers of the release reaction and hence further platelet aggregation. The prevention of continuing aggregation of platelets by aspirin may contribute to its tendency to cause gastrointestinal bleeding. Dipyridamole inhibits platelet phosphodiesterase, thus leading to elevated levels of platelet cyclic AMP and reducing platelet aggregation. It may also act by inhibiting prostaglandin production in platelets. Dextran, a branched polysaccharide, decreases the incidence of arterial thrombi but the mechanism of its action is not understood.

*Anticoagulant Therapy.* Anticoagulants interfere with fibrin formation, but have no effect on platelet function. They are used in the prevention and treatment of venous thrombosis. The two most widely-used anticoagulants are warfarin and heparin; there are in addition others, such as ancrod, which show promise for the treatment of venous thrombosis.

Warfarin acts by preventing the production of those functional coagulation factors which require vitamin K for their biosynthesis. These are factors VII, IX, X and prothrombin. In each case the protein is synthesised by membrane-bound ribosomes, and is modified by a vitamin K-dependent carboxylation of specific glutamic acid residues to yield γ-carboxyglutamate. Warfarin interferes with the action of this coenzyme (Figure 9.6), preventing the production of the γ-carboxyglutamate residues. These residues are required for the interaction of vitamin K-dependent clotting factors with the platelet phospholipid surface via $Ca^{2+}$ bridges (see Figure 9.2). Such interactions will be decreased in patients undergoing warfarin therapy and thus the activated factors will not be generated. Thrombin, for example, is produced at a significant rate only when the complex of prothrombin, factor Xa, $Ca^{2+}$, platelet phospholipid and factor V is formed. Since warfarin acts by interfering with the biosynthesis of functional prothrombin, which has a half-life of about 48 hours, it takes 48–72 hours before warfarin has a significant anticoagulant effect.

**Figure 9.6: Inhibition of Carboxylation of Prothrombin Precursor by Warfarin**

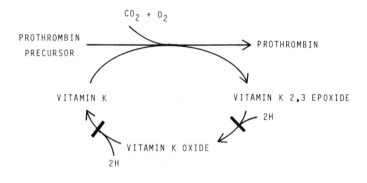

Heparin is a naturally-occurring acidic polysaccharide. This highly-sulphated polymer consists of repeating disaccharide units containing uronic acid (iduronic or glucuronic acid) and $N$-acetylglucosamine. The sulphate groups are attached to either the uronic acid or amino sugar moieties. The mechanism by which heparin acts as an anticoagulant *in vivo* is discussed on p. 123.

There are several situations in which heparin is administered as an anticoagulant. One use is in conjunction with warfarin during the 48-hour period when warfarin is not fully effective. Heparin is used also in haomodialyoio and cardiac surgery wheie theie is exliacoipoieal circulation. Recently 'low dose heparin' therapy has been introduced. This involves subcutaneous injections of amounts of heparin lower than those required to affect clotting times as measured *in vitro*. Nevertheless, this treatment has a significant antithrombotic effect in cases of deep vein thrombosis.

Ancrod is a proteolytic enzyme isolated from the venom of the Malayan pit viper (*Ankistrodon rhodostoma*). Ancrod is a thrombin-like proteinase having a more limited action than thrombin itself. It cleaves fibrinopeptide A from fibrinogen but is incapable of performing some of the other reactions of thrombin, namely releasing fibrinopeptide B, activating factor XIII or promoting platelet aggregation. The release of fibrinopeptide A from fibrinogen produces a form of fibrin with little mechanical strength which is rapidly degraded by the fibrinolytic system. The levels of other coagulation factors and platelets are unaffected. Thus, the blood of patients treated with ancrod is still capable of generating thrombin but it cannot clot because insufficient fibrinogen is present. A further beneficial effect of this treatment is that the reduction of fibrinogen levels also encourages blood flow.

*Thrombolytic Therapy.* Thrombolytic agents are those which increase the rate of fibrin dissolution, either directly or indirectly. Whereas anticoagulants are designed to prevent thrombosis and enlargement of pre-existing thrombi, thrombolytic agents are designed to promote the degradation of thrombi. Streptokinase, isolated from haemolytic Streptococci, is the most widely-used agent of this type; urokinase, purified from male urine, is another example. In spite of its name, streptokinase is a protein with no known enzymic activity. It combines in a 1:1 molar complex with plasminogen, which then undergoes a conformational change as a result of the interaction. This complex activates other plasminogen molecules leading to the rapid generation of plasmin. Plasmin bound to fibrin leads to dissolution of the thrombus.

Urokinase is also an activator of plasminogen. In this case plasmin is produced by cleaving two peptide bonds in plasminogen. Thus, both streptokinase and urokinase act by triggering the natural thrombolytic system.

Each of the various anticoagulant and thrombolytic agents described has certain advantages and disadvantages. Warfarin is the only agent which can be administered orally. Others must be given intravenously, because of their high molecular weights and susceptibility to degradation in the gastrointestinal tract. Warfarin, heparin and urokinase are not antigenic, warfarin because it has a low molecular weight, and heparin and urokinase because they are normal body constituents. Both ancrod and streptokinase are antigenic and prolonged administration can lead to resistance.

## Further Reading

Bailey, J.M. (1979) 'Prostacyclins, Thromboxanes and Cardiovascular Disease', *Trends in Biochemical Sciences, 4*, 68–71

Gaffney, P.J. (1981) 'Thrombosis: A Molecular Approach to Therapy', *Nature, 290*, 445–6

Heinmark, R.L., Kurachi, K., Fujikawa, K. and Davie, E.W. (1980) 'Surface Activation of Blood Coagulation, Fibrinolysis and Kinin Formation', *Nature, 286*, 456–60

Jackson, C.M. and Nemenson, Y. (1980) 'Blood Coagulation', *Annual Reviews of Biochemistry, 49*, 765–811

Mustard, J.F., Kinlough-Rathbone, R.L. and Packham, M.A. (1980) 'Prostaglandins and Platelets', *Annual Review of Medicine, 31*, 89–96

Rákóczi, I., Wiman, B. and Collen, D. (1978) 'On the Biological Significance of the Specific Interaction between Fibrin, Plasminogen and Antiplasmin', *Biochimica et Biophysica Acta, 540*, 295–300

Zucker, M.B. (1980) 'The Functioning of Blood Platelets', *Scientific American, 242*, 70–89

# 10   THE BLOOD BUFFERING SYSTEMS

## Introduction

The six litres (l) of blood in the circulation is in contact with the cytoplasm of the tissues indirectly via the much larger volume (30 l approximately) of extracellular fluid, which bathes the cells. The blood not only supplies the tissues with nutrients and oxygen, but also receives metabolites and waste products, including $CO_2$, from them. Thus, the maintenance of homeostasis in the tissues is dependent upon control of the composition of blood between narrow limits. This is achieved by the lungs, which remove $CO_2$ from the blood, while the kidneys control the $HCO_3^-$ and other ion concentrations and the pH of the blood. The kidneys filter the blood, generating approximately 150 l of fluid each day; most of this is reabsorbed and a concentrated urine (approximately 1.5 l) is produced. The fact that urine volume can be varied independently of solute concentration is very important in terms of the regulation of osmolality of the blood and its total volume. However, a detailed discussion of renal function is outside the scope of this book.

This chapter describes how short-term changes in pH are counteracted by the blood's buffering system, which is comprised of bicarbonate, haemoglobin and plasma proteins. In addition, those movements of ions across the red cell membrane that are associated with the transport of $CO_2$, and their relationship to the control of blood pH, are discussed.

## Ionic Composition of Blood

The ionic composition of blood cells differs considerably from that of plasma. From the point of view of blood buffering the red cells are more important than white cells or platelets. Accordingly, the ion content of erythrocytes only will be considered, and a comparison of their ionic composition with that of plasma is shown in Table 10.1. Erythrocytes, like other cells, contain a high $K^+$ concentration (required for glycolysis) but low concentrations of $Na^+$ and $Cl^-$. Most anions, including $HCO_3^-$ and $Cl^-$, cross the membrane by facilitated diffusion on the anion exchange protein (see p. 9); this results in the red cell

131

membrane being $10^6$-$10^7$ times more permeable to anions than to cations. The distribution of $Na^+$ and $K^+$ across the membrane is maintained by the $Na^+$, $K^+$-ATPase. For each ATP molecule hydrolysed, the ATPase transports $3Na^+$ out of, and $2K^+$ into, the red cell. $Ca^{2+}$ is transported out of red cells by the $Ca^{2+}$-ATPase, and in old cells the decline in activity of this enzyme leads to a rise in intracellular $Ca^{2+}$, that triggers a sequence of events leading to red cell destruction in the spleen. Thus, in contrast to anion transport, the distribution of cations across the membrane is maintained by an active transport process. Haemoglobin bears a net negative charge and represents a large pool of fixed negative charge trapped inside the red cell. It is the distribution of cations, together with the fact that the negatively-charged haemoglobin is confined within the cell, that governs the distribution of the diffusible anions (mainly $HCO_3^-$ and $Cl^-$) across the membrane according to the Donnan equilibrium. The net result of these ion movements is the creation of ion concentration differences, which lead to the separation of charge across the membrane. The ions will tend to move down their electrochemical gradients and this movement at the membrane surface forms the basis of the membrane potential (negative inside; positive outside).

**Table 10.1: A Comparison of the Inorganic Ion Composition of Red Cells and Plasma**

| Ion | Red cell (mmol per litre of packed cells) | Plasma (mmol per litre) |
|---|---|---|
| $Na^+$ | 19 | 140 |
| $K^+$ | 136 | 4 |
| $Cl^-$ | 78 | 100 |
| $Mg^{2+}$ | 3 | 1 |
| $Ca^{2+}$ | 0.001 | 2.5 |
| $HCO_3^-$ | 11 | 27 |
| Inorganic phosphate | 0.4 | 1 |
| Protein | 340 mg[a] | 78 |

Note: a. Mainly haemoglobin.

### Carbon Dioxide Transport in the Blood

Most of the $CO_2$ produced by metabolism in the tissues diffuses into the blood. It passes into the red cells where some combines with haemoglobin to form carbaminohaemoglobin (see Chapter 3). The remainder is converted to carbonic acid under the influence of the red-cell enzyme, carbonic anhydrase. The carbonic acid so formed

**Figure 10.1: Transport of Carbon Dioxide and the Bicarbonate Buffering System in Blood**

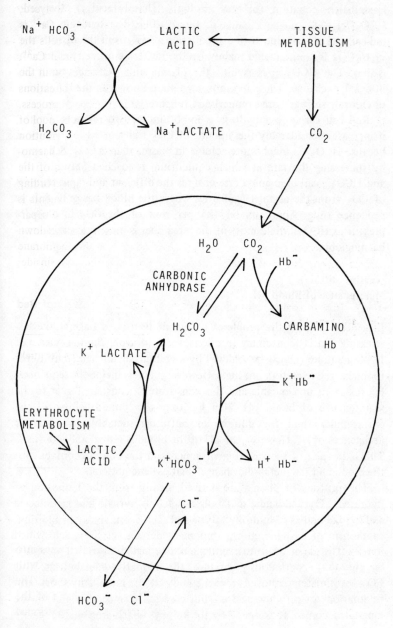

dissociates into $H^+$ and $HCO_3^-$ (Figure 10.1). Knowing the pH of the blood and the p$K$ value of the dissociation, and by using the Henderson-Hasselbalch equation (pH=p$K$ + $\log_{10}$ {[base]/[acid]}), the ratio $HCO_3^-/H_2CO_3$ can be calculated: this is 20:1. The tendency for the red-cell pH to fall due to the production of $H^+$ from the dissociation of $H_2CO_3$ is counteracted mainly by the buffering capacity of haemoglobin. The $HCO_3^-$ passes into the plasma and exchanges with $Cl^-$ in a 1:1 exchange. These ion movements are known as the Hamburger or chloride shift and are summarised in Figure 10.1.

Red cell carbonic anhydrase is important in two respects. Firstly, it increases considerably the amount of $CO_2$ transported in the blood, because $HCO_3^-$ is much more soluble in plasma than is $CO_2$. Secondly, by increasing the rate at which equilibrium is reached between $CO_2$ and $HCO_3^-$, carbonic anhydrase ensures the efficient and rapid removal of $CO_2$ from the alveolar capillaries where the blood has only a short residence time. Approximately 85 per cent of the $CO_2$ in blood is present as $HCO_3^-$ while most of the remainder is present as carbamino-haemoglobin.

## Maintenance of Blood pH

In general, one of the problems of metabolism is to balance the $H^+$-releasing and $H^+$-consuming reactions, so as to avoid the net release of $H^+$. Sometimes this can be achieved by metabolising uncharged nutrients to uncharged excretory products. However, in most metabolic sequences, $H^+$ release is unavoidable and this constitutes a constant threat to the maintenance of blood pH. Man has only the limited volume of the extracellular fluid for diluting or buffering metabolic excesses or deficiences of $H^+$. However, the pH of the blood is remarkably constant. This is in spite of the varied environments in the body in contact with the blood and the metabolic changes that are encountered.

The protons in plasma are derived mainly from the ionisation of hydrated $CO_2$ and acidic metabolites, such as pyruvic and lactic acids and ketone bodies ($\beta$-hydroxybutyric and acetoacetic acids). In addition, catabolism of cysteine and methionine generates sulphuric acid which ionises. The major threat to maintenance of constant pH is that presented by the $CO_2$ continuously produced by oxidative metabolism. This $CO_2$ is ultimately removed from the body by the lungs. Any short-term fluctuations of carbonic acid are buffered by haemoglobin and, in the long-term, control is exercised by the kidneys which are able to regulate blood bicarbonate levels. Strong acids, like $H_2SO_4$, and organic acids stronger than carbonic acid, are buffered in the short-term by the major plasma buffer, bicarbonate.

The first problem of $CO_2$ transport is to make tolerable the extra hydrogen ion added to the blood as it acquires a $CO_2$ molecule during its passage through the tissue capillaries (see Figure 10.1) and yet keep the $H^+$ available for regenerating the $CO_2$ when released in the lungs. Since such a large amount of extra $HCO_3^-$ is in fact added to the blood when $CO_2$ enters, it is essential that an equivalent amount of $H^+$ is bound if the pH is to remain unaltered. Most of the buffering of this $H^+$ from carbonic acid in blood is due to haemoglobin (Figure 10.1). It can be calculated that simple buffering by oxyhaemoglobin will normally allow the blood to take up and release $CO_2$ without a pH change of more than approximately 0.1 of a pH unit. However, measurement of arterial and venous pH shows that, at rest, they differ by no more than 0.03 of a pH unit. Hence, less than half of the ordinary buffering activity of haemoglobin appears to be used. The implication is that hydrogen ions released on the uptake of $CO_2$ are mainly disposed of in some other way. This is explained by the fact that as $CO_2$ is entering the blood, oxygen is leaving the blood and as a result, about 30 per cent of the oxyhaemoglobin is changed to deoxyhaemoglobin. When oxyhaemoglobin loses $O_2$ forming deoxyhaemoglobin the affinity of the protein for $H^+$ increases, as discussed above. Thus, because part of the oxyhaemoglobin is converted to deoxyhaemoglobin (occasioned largely by the fall in $pO_2$), a major part of the uptake of $H^+$ by haemoglobin occurs independently of the pH fall. At the pulmonary end of the cycle $CO_2$ is readily regenerated because the $H^+$ necessary for the reaction:

$$HCO_3^- + H^+ \rightleftharpoons CO_2 + H_2O$$

is again made available from deoxyhaemoglobin. More becomes available as its release from oxyhaemoglobin is stimulated by the entering $O_2$ which converts deoxyhaemoglobin to oxyhaemoglobin.

Although it is clear that haemoglobin forms a vital part of the blood buffering system with respect to the transport of $CO_2$ and the carriage of $H^+$, this mechanism cannot function in buffering the $H^+$ produced during intermediary metabolism. Other blood buffers are involved, namely bicarbonate, phosphate and plasma proteins. The relative strengths of the acids involved must be considered. Compounds (or mixtures of compounds) are buffers if they form weak acids when stronger acids are added to them, since the $H^+$ concentration is decreased effectively by the lower ionisation of the weaker acid. For example, if HCl (a strong acid) is added to sodium acetate, sodium chloride and acetic acid (a weak acid) are formed. As far as the buffering capacity of blood is concerned, it is important that haemoglobin is a weaker acid than $H_2CO_3$, which in turn is a weaker acid than sulphuric, phosphoric, pyruvic or lactic acid. For instance, in the tissues, lactic acid enters the blood where it is buffered by $HCO_3^-$ which forms $H_2CO_3$. The $H_2CO_3$ enters the red cell and is, in turn, buffered by haemoglobin. Similarly, lactic acid is generated by the erythrocyte's own metabolism, and the lactate formed from this will also be buffered by haemoglobin. Thus, the relative strengths of the blood buffers form a system (shown in Figure 10.1) that can counteract short-term pH changes. In the lungs, when $CO_2$ is removed, the events occurring inside the red cell in Figure 10.1 are reversed.

## Further Reading

Davson, H. and Segal, M.B. (1976) 'Homeostasis of the Blood Composition' in H. Davson and M.B. Segal (eds), *Introduction to Physiology, Volume 3*, Ch. 5, pp. 404–575, Academic Press Inc. (London) Ltd, London

Montgomery, R., Dryer, R.L., Conway, T.W. and Spector, A.A. (1980) 'Acid-Base, Fluid, and Electrolyte Control' in R. Montgomery, R.L. Dryer, T.W. Conway and A.A. Spector (eds), *Biochemistry. A Case-oriented Approach*, Ch. 4, 3rd edn, The C.V. Mosby Company, St Louis, Missouri

# 11    BLOOD AS A TRANSPORT SYSTEM

## Introduction

Blood travels to all parts of the body, and the flow rate in most parts of the circulation is rapid. It therefore provides an efficient transport system for many essential nutrients, waste substances and effector molecules (e.g. hormones). By providing a link between separate tissues, blood enables the biochemical activities of one organ to influence the activities of another. Nutrients absorbed from the gastrointestinal tract are transported by the blood to the tissues, metabolites are moved between organs and waste substances are transported for excretion; hormones produced by one organ are carried in the blood plasma to their target tissues. In this way the transport functions of blood play an essential role in the overall homeostatic mechanisms of the body by co-ordinating the biochemical activities of its different parts. In addition, several blood components bind and transport foreign substances (e.g. drugs, antibiotics) introduced into the body and can markedly influence their pharmacological or toxicological effects.

This chapter deals with the function of blood as a transport system, but excludes the carriage of $O_2$ and $CO_2$, which was dealt with in Chapters 3 and 10.

## Transport Mechanisms in Blood

A wide range of substances, both naturally-occurring and synthetic, are transported in the plasma in three ways: in multimolecular complexes; bound to a plasma protein; and in free solution:

(1) The plasma lipids, triglycerides and cholesterol, are transported in lipoprotein particles. These are composed of a core of triglyceride and cholesterol (mainly as cholesterol ester) with an outer layer of proteins (apoproteins), phospholipid and some cholesterol. The incorporation of water-insoluble lipids, such as triglyceride and cholesterol ester, into lipoproteins stabilises these lipids in plasma and facilitates their transport between tissues.

(2) Other water-insoluble, as well as some water-soluble, compounds

137

may be bound to plasma proteins. By far the most important protein in this respect is albumin; one of its main functions is the transport of fatty acids, but it also binds many other substances, including bilirubin, haem, heavy metal ions and some hormones, as well as many drugs and antibiotics. There are also several other more specific carrier proteins, including those which bind steroid hormones, haem or some vitamins. Binding to carrier proteins has several advantages. It may increase the amounts of a substance that can be carried – for example iron transport by transferrin. It allows transport of relatively insoluble molecules in the blood, whilst the rate of excretion of soluble molecules may be lowered by reducing their glomerular filtration. In addition, the protein-bound molecules may be more readily targeted to receptors in particular tissues for uptake into cells – for example erythroblasts have transferrin receptors.

(3) Water-soluble substances are also carried in true solution in the plasma. Included in this category are urea, many ions, simple metabolites like glucose and lactate, and polypeptide hormones such as insulin and glucagon.

## Transport of Lipids by Lipoproteins

The lipoproteins are heterogeneous both in size and composition but certain classes of particle can be distinguished on the basis of either density or net charge (see Table 11.1). Triglycerides (0.9 g/ml) are less dense than cholesterol and phospholipids (1.0 g/ml), which are in turn less dense than protein (1.3 g/ml). Because lipoprotein classes contain not only different types of lipid but also different relative amounts of lipid and protein, they can be separated by ultra-centifugation on density gradients. This produces four major classes of lipo-protein, namely chylomicrons, very low density lipoproteins (VLDL), low density lipoproteins (LDL) and high density lipoproteins (HDL) (see Table 11.1). Lipoproteins can also be separated on the basis of differences in their net charge, arising mainly from differences in the type of apoproteins they contain (although phospholipids also contrib-ute to surface charge). The apoproteins are of different types (apo A-E) and are characteristically distributed amongst the lipoprotein classes (Table 11.1). In addition, apoA and apoC consist of several sub-types (apoC-I, apoC-II, etc.), some of which have specific biochemical functions; for instance, apoA-I activates the plasma enzyme lecithin-cholesterol acyltransferase (LCAT), whilst apoC-II activates the enzyme lipoprotein lipase. Despite their structural heterogeneity, each of the lipoprotein classes has an identifiable role (Table 11.1). The main

**Table 11.1: Plasma Lipoproteins: Their Composition, Properties and Functions**

| Lipoprotein class | Chylomicrons | VLDL | LDL | HDL |
|---|---|---|---|---|
| Density range (g/ml) | 0.92–0.96 | 0.96–1.01 | 1.01–1.06 | 1.06–1.21 |
| Size (nm) | 30–500 | 30–100 | 20–25 | 10–15 |
| Electrophoretic mobility | Remain at origin | pre-$\beta$-globulin | $\beta$-globulin | $\alpha$-globulin |
| Composition (% wt) | | | | |
| Cholesterol | 8 | 22 | 46 | 30 |
| Phospholipid | 7 | 18 | 22 | 29 |
| Triglyceride | 83 | 50 | 10 | 8 |
| Apoprotein | 2 | 9 | 21 | 33 |
| Major apoproteins | apo B, C, A₁ | apo B, C, E | apoB | apoA |
| Major function | Triglycerides from gastrointestinal tract to tissues | Triglycerides from liver to tissues | Cholesterol ester to tissues | Cholesterol ester to liver |

function of chylomicrons and VLDL is the transport of triglycerides, while LDL and HDL transport cholesterol, largely as cholesterol ester.

Plasma triglyceride represents a major energy source, and up to 150 g of triglyceride are transported daily in the blood and metabolised or stored by the tissues. Because there is a rapid flux of triglyceride through the blood stream, the maximum plasma pool of triglyceride in lipoproteins is only 2–5 g. Each day 50–100 g of dietary triglyceride is transported from the intestine to the tissues in the form of chylomicrons. These are assembled inside the mucosal cells of the intestine, enter the lymphatic system and discharge into the blood stream through the thoracic duct. After a very fatty meal the blood has a 'milky' appearance due to light scattering as a result of the size and large number of chylomicrons present. The peripheral tissues sequester the chylomicrons and their constituent triglyceride is hydrolysed to fatty acids and monoglyceride by the enzyme lipoprotein lipase (LPL) bound on the lumenal surface of blood-capillary endothelial cells. This is a rapid process, most of the chylomicrons being removed from the circulation within minutes after entry. LPL is activated by apoC-II, which the chylomicrons receive by exchange, probably from HDL, once they enter the blood stream. Most of the fatty acids released by LPL are taken up and utilised by the tissues, whilst the remainder are bound by albumin in the plasma (see below). This process of triglyceride hydrolysis leaves a so-called chylomicron 'remnant', which is probably taken up by the liver and catabolised.

The other way in which triglyceride is transported in the blood is in VLDL, most of which are synthesised and assembled in the liver using fatty acids either taken up from circulating albumin or synthesised *de novo* in the liver cells. Hepatic VLDL contributes 20–50 g of triglyceride each day to the circulation. Once in the blood the VLDL triglyceride is metabolised within an hour or so. The triglyceride is hydrolysed by LPL in extrahepatic tissues; LPL is activated by apoC-II as described above. In adipose tissue the released fatty acids are used to resynthesise triglyceride which is either stored or used to provide energy. In muscle the fatty acids are oxidised as a source of energy.

During the hydrolysis of triglyceride in VLDL, the apoB is retained by the lipoprotein but other apoproteins are lost. Together with the exchange of components between lipoproteins and the transfer of cholesterol ester from HDL, triglyceride hydrolysis leads to the stoichiometric formation of one LDL particle from one VLDL particle. Thus, LDL differ from other lipoprotein classes in that they are not assembled within a tissue, but are formed within the blood as a result of metabolism of another lipoprotein class.

LDL represent the major transport form of cholesterol. The cholesterol in LDL is metabolised more slowly ($t_{\frac{1}{2}} \simeq$ 3-4 days) than is the triglyceride in VLDL. In the peripheral circulation LDL are bound to cell-surface receptors and taken up into cells by endocytosis. The uptake of LDL, and thus cholesterol, is regulated by the number of LDL receptors. In one form of *familial hypercholesterolaemia* the normal uptake of LDL is impaired, leading to elevated levels of plasma cholesterol. Once inside cells the cholesterol ester of LDL is converted to cholesterol by a specific lysosomal acid lipase. The cholesterol is either used for membrane biosynthesis or is re-esterified and stored within the cells.

In *Wolman's disease* the acid lipase is absent, which leads to the over-accumulation of cholesterol ester inside cells and normally results in death during infancy. Cholesterol may also be synthesised *de novo* from acetate; the amount of this endogenous synthesis is controlled by the feedback inhibition of 3-hydroxy-3-methylglutaryl CoA reductase by cholesterol.

The other way in which cholesterol is transported in the plasma is in HDL. There are different types of HDL particle, depending upon their site of synthesis and assembly. Most HDL are assembled by the intestinal mucosa using cholesterol. Another type is produced by the liver using cholesterol synthesised *de novo*. The HDL are secreted by the liver as 'nascent' HDL; once in the circulation they acquire components, that are released from VLDL during their conversion into LDL, to become mature HDL particles. These HDL may pick up excess cholesterol from the tissues and transport it back to the liver for excretion in the bile. Some cholesterol is taken up from HDL (and LDL) into the gonads and adrenal cortex, which convert it into steroid hormones.

The levels of circulating cholesterol, particularly that contained in LDL, may be important in the promotion of vascular disease such as atherosclerosis. Whatever the cause of atherosclerosis, the major feature of the atheromatous plaque is a massive accumulation of cholesterol between and within the smooth-muscle cells of the intima and media. It may be significant that circulating levels of LDL increase in people at an age when susceptibility to atherosclerosis also rises. There are no drugs that reliably reduce circulating cholesterol levels. Clofibrate is commonly used, for instance, after infarction, but the way in which it lowers plasma LDL levels is not known.

There are various disorders of lipoprotein metabolism resulting in increased or, less frequently, decreased levels of lipoproteins. These may be categorised as being primary (hereditary), secondary (associated with a disease; for example, diabetes mellitus, hyperthyroidism, biliary obstruction or pancreatitis) or acquired, for example, in overweight persons or induced by diet (including alcoholism). The various types of *hyperlipoproteinaemia* are classified as types I-V according to the class of lipoprotein that is elevated. For instance, in Type IIa only LDL are raised, whilst in Type IIb both LDL and VLDL are raised. Because of the different lipid compositions of LDL and VLDL, the two types of hyperlipoproteinaemia can be distinguished by measurements of blood lipids; in Type IIa the level of cholesterol is elevated but triglyceride is normal, whilst in Type IIb the levels of both cholesterol and triglyceride are elevated. There may also be associated changes in red cells, because erythrocytes exchange lipids with lipoproteins directly and via the action of LCAT. The cholesterol content and phospholipid composition of red cell membranes may alter, sometimes leading to morphological changes (see Chapter 2). Type I hyperlipoproteinaemia differs from other types in being caused by an autosomal recessive deficiency of LPL, which leads to high plasma levels of chylomicrons and, therefore, triglyceride.

*Hypolipoproteinaemias* are much rarer. In the recessively inherited *Tangier disease* there is a deficiency of HDL, so that cholesterol ester transport is impaired, leading to its accumulation in the reticuloendothelial system, particularly in the liver, spleen and tonsils (which become bright yellow).

### Transport on Albumin and Other Binding-proteins

Albumin is the most abundant protein in plasma (40-50 g/l; approximately 50 per cent of the total). It possesses a large number of ionisable groups and is consequently very soluble. There are two main functions of albumin. Firstly, it exerts 80 per cent of the colloid osmotic pressure of blood and is thus an essential factor in the maintenance of blood volume. Secondly, it is a transporter of many substances, most notably organic anions such as fatty acids and bilirubin, the high solubility of albumin providing a large transport capacity.

Albumin is synthesised by the liver, 8-14 g each day and 6-10 per cent of that in the blood is metabolised daily. It is not known where breakdown occurs nor how it is controlled. Albumin synthesis is decreased in *cirrhosis*, due to the reduced synthetic capacity of the liver, and in *Kwashiokor*, due to a shortage of amino acid precursors. The lowered amount of blood albumin reduces the colloid osmotic pressure of the blood leading to oedema in the tissues; there may also be disturbances in lipid metabolism.

Albumin is a protein of molecular weight 69,000, the amino acid sequence of which is known. Its single polypeptide chain is folded to produce a three-dimensional structure consisting of three repeating units (domains). Each domain is composed of six peptide loops stabilised by disulphide bridges and there is much sequence homology between the three domains. One domain has two high affinity binding sites for fatty acids, while the other domains have low affinity sites. Long chain fatty acids, such as palmitic and linoleic, are insoluble in plasma, but by binding to albumin they are rendered water-soluble. The amount of fatty acid bound to plasma albumin, even if both sites are filled, is comparatively small (less than 0.3 g/l of plasma) compared with triglyceride fatty acids in lipoproteins. However, this bound fatty acid is turned over rapidly, *viz.* 99 per cent within ten minutes, and is capable of supplying many extra-hepatic tissues with more than half their energy requirements.

Albumin also has the capacity to bind many substances, both naturally-occurring and synthetic. It binds bilirubin, haem and haemin, which are responsible for its slightly yellow colour. Haem is also bound to a specific glycoprotein, *haemopexin*. The haem–haemopexin complex is taken up rapidly by the liver, which converts the haem to bile pigments. Bilirubin is transported on albumin to the liver for conjugation and subsequent excretion (see Chapter 4); free bilirubin is toxic, but when bound to albumin it is rendered non-toxic. In the new-born the bilirubin-binding capacity of albumin may be exceeded, because the liver in such infants has a low capacity for excreting bilirubin in bile (a task that is performed for the fetus by the placenta). The presence of free bilirubin in plasma causes jaundice which, if severe, may lead to brain damage (*kernicterus*), because the lipophilic bilirubin crosses the blood–brain barrier and is concentrated in the brain.

Albumin binds steroid hormones and thyroxine, although other plasma glycoproteins also carry out this function. For instance, there is a specific *thyroxine-binding protein*, while *transcortin* binds and transports steroid hormones such as oestradiol, progesterone and testosterone.

Many biologically important metal ions bind to albumin. However, copper and iron have specific transport mechanisms. Copper is bound by *caeruloplasmin*, 6–8 molecules of copper per molecule of caeruloplasmin. One of its functions is to transport copper to the liver, which regulates copper levels by excreting any excess in the bile. In *Wilson's disease* copper accumulates in the liver although most patients have a deficiency of caeruloplasmin. Iron is bound tightly and transported by *transferrin*, two molecules of Fe(III) per molecule of transferrin. Like caeruloplasmin, transferrin is synthesised in the liver. Transferrin transports iron from the gastrointestinal tract and liver to the bone marrow for haemoglobin synthesis during erythropoiesis (see Chapter 5).

Besides the physiological compounds already described, albumin is capable of binding a very wide range of synthetic compounds, such as antibiotics and drugs, that may be present in plasma. The binding of drugs to albumin and other plasma proteins is an important factor in considerations of drug potency and pharmacological activity. Both the extent and the strength of binding are relevant to the absorption, distribution, metabolism and excretion of drugs. For example, a high degree of plasma-protein binding could substantially enhance the absorption of a drug from the gastrointestinal tract by producing a favourable concentration gradient across the mucosal cells. Protein-bound drugs are normally considered to be biologically inactive and thus, although their potential activity is suppressed, they form a reservoir from which active drug is released. For instance, the large number of kinds of sulphonamide antibiotics can be broadly divided into two groups. The first group has a lipid-soluble side-chain, which binds to albumin and is delivered slowly to the tissues; the second group has a side-chain that suppresses the ionisation of the sulphone group, which leads to their rapid reabsorption by the kidneys so they are useful in the treatment of urinogenital tract infections. Many semi-synthetic penicillins, for example, oxacillin and flucloxacillin, are largely bound (80–95 per cent) to albumin in the circulation; ampicillin is notable for being poorly bound (approximately 20 per cent).

Although binding to plasma proteins affects the availability of a drug it may not decrease the rate of clearance of all drugs from the plasma. It has been claimed that when some drugs are bound to proteins they are taken into cells by pinocytosis. In this case their uptake by extravascular tissues would be enhanced.

The very wide range of structures of drugs and antibiotics makes it difficult to generalise, but a number of basic features of drug binding to plasma proteins are worth noting. Firstly, drugs having similar chemical characteristics to endogenous compounds that are normally bound to plasma proteins will be similarly bound. For example, aspirin displaces bilirubin from albumin. Secondly, it is important to recognise that drug binding may substantially alter the binding capacity for endogenous materials. Conversely, increased levels of endogenous materials may lead to changes in drug binding. In short, there may be competition for binding between endogenous and exogenous compounds.

Albumin is quantitatively the most important protein that binds drugs and drug-transporting capacity is related to albumin levels. Compared with the normal healthy adult, the levels of serum albumin are lower in infancy, pregnancy, senility and in various disease states. In addition, stress, injury and alcohol decrease albumin synthesis. A reduction in serum albumin leads to a significant increase in free drug levels with attendant enhancement of therapeutic or toxicological effects.

## Metabolite Transport and Inter-organ Metabolism

The differentiated tissues of the body possess particular functions and this functional specialisation is reflected in their metablism. In some instances the primary function of a tissue is metabolic and its role is to keep other tissues provided with essential metabolites. Adipose tissue, which is specialised for the synthesis, storage and mobilisation of triglycerides, is an example of such a tissue. In contrast, some tissues, for example brain, may make very little contribution to the provision of metabolic fuels but can place considerable demands on other tissues to supply metabolites. To achieve this metabolic co-operation between tissues, a transport system is required: a role served by the blood.

There are a number of metabolites present in the blood concerned with energy metabolism. These include glucose, lactate/pyruvate, glycerol, triglycerides, fatty acids, ketone bodies and certain amino acids, particularly alanine and glutamine. The relative importance of these fuels depends on many factors such as the diet, exercise, state of fasting and hormonal balance. Glucose is often regarded as a major metabolite because of the reliance of the brain on this carbohydrate. Quantitatively, however, fats are more important to most tissues in the body. In addition to releasing fatty acids, the mobilisation of triglycerides in adipocytes releases glycerol, which is transported in the blood to the liver where it serves as a gluconeogenic precursor. The other main gluconeogenic precursors being transported to the liver are: (a) lactate/pyruvate released from erythrocytes and white skeletal muscle during contraction; and (b) amino acids, notably alanine derived from the catabolism of muscle protein. The glucose produced by hepatic gluconeogenesis during fasting is used by the erythrocytes or the brain. Because the end product of erythrocyte metabolism is lactate, the metabolisms of liver and erythrocytes create a closed loop. Unlike erythrocytes, brain completely oxidises glucose to $CO_2$.

The ketone bodies, acetoacetate and $\beta$-hydroxybutyrate, are a metabolic fuel that normally is present in the blood in very low amounts; but during fasting the liver synthesises ketone bodies from acetyl-CoA obtained by the $\beta$-oxidation of fatty acids. These ketone bodies leave the liver and are transported to other tissues, such as muscle, for further oxidation. In this way the metabolism of fatty acids mobilised from adipose tissue is divided between the liver and extrahepatic tissues. If the fasting period extends beyond about seven days, the metabolism of the brain adapts with part of the energy requirement being met from ketone body oxidation rather than glucose consumption. This adaptation helps to conserve the body's protein reserves.

## Further Reading

Anon. (1980) 'Lipoprotein Structure', *Annals of the New York Academy of Sciences, 348*, 1–436

Berde, C.B., Hudson, B.S., Simoni, R.D. and Sklar, L.A. (1979) 'Human Serum Albumin. Spectroscopic Studies of Binding and Proximity Relationships for Fatty Acids and Bilirubin', *Journal of Biological Chemistry, 254*, 391–400

Blombäck, B. and Hanson, L.A. (1979) *Plasma Proteins*, John Wiley & Sons Ltd, Chichester

Brown, M.S., Kovenen, P.T. and Goldstein, J.L. (1981) 'Regulation of Plasma Cholesterol by Lipoprotein Receptors', *Science, 212*, 628–35

Gillette, J.R. (1973) 'Overview of Drug–Protein Binding', *Annals of the New York Academy of Sciences, 266*, 6–17

Kragh-Hansen, U. (1981) 'Molecular Aspects of Ligand Binding to Serum Albumin', *Pharmacological Reviews, 31*, 17–53

# 12 THE COLLECTION, SEPARATION AND ANALYSIS OF BLOOD AND ITS COMPONENTS

**Blood Collection and Fractionation**

The widespread use of *blood transfusion* is taken for granted nowadays and blood donor sessions are regular features of the life of many communities. The key to successful transfusion was the discovery of the major blood groups (see Chapter 8) by Landsteiner in 1900-1, and the first transfusions were made during the first World War.

Nowadays, in many countries, donated whole blood is collected by a blood transfusion service and distributed via hospital blood banks. Whole blood is most commonly used for transfusion but red cells, platelets, plasma and plasma protein fractions are also used and they are prepared from collected blood; placentas may serve as a source of some plasma protein fractions. Although blood transfusion services do prepare certain plasma protein fractions, the majority are isolated by commercial organisations.

A blood donor gives approximately 450 ml of blood (termed one 'unit') per session. It is collected into approximately 65 ml of a buffered solution containing citrate and glucose. The two collection media in common use are ACD (acid–citrate–dextrose) and CPD (citrate–phosphate–dextrose). The citrate serves as an anticoagulant by chelating calcium ions, while glucose is required as an energy source for the erythrocytes. The glucose content of donated blood would be exhausted within approximately nine days at $4°C$, but the added glucose enables the blood to be stored for a further eight weeks at this temperature. Although some blood is kept frozen, most is stored at $4°C$ and is used within one week of collection. The maximum shelf-life is set at three weeks, after which time more than 30 per cent of the cells are non-viable. These 30 per cent would be removed from the circulation of a transfused patient within 24 hours. Cells which survive this initial period have normal life-spans.

The rate of glycolysis in blood stored at $4°C$ decreases dramatically thereby lowering ATP levels and slowing the ion pumps. This results in a progressive reduction of the $Na^+$ and $K^+$ gradients across the membrane. The end-products of glycolysis, lactate and pyruvate, accumulate and lower the red cell pH, which further inhibits glycolysis. These effects lead to a progressive decrease in the concentrations of ATP and 2,3-DPG; for instance, 2,3-DPG concentration falls to 20 per cent of normal after one week of storage at $4°C$, increasing the blood's affinity for oxygen and impairing its release in the tissues. However these changes are probably not important clinically, unless massive transfusion is undertaken, when fresh blood should be used. In Sweden and West Germany a further additive used in adenine, which reduces the rate of depletion of ATP and 2,3-DPG. Adenine probably acts by maintaining the adenine nucleotide pool and such blood can be stored for five to six weeks.

An alternative storage method is to separate the cells from the plasma and store both frozen at $-80°C$ or $-150°C$. Haemolysis of the cells during freezing is prevented by adding a cryoprotective agent such as glycerol. Once thawed the cells must be washed free of glycerol before use. The main advantage of using frozen cells is that their shelf-life is greater than a year, because the biochemical changes seen in blood stored at $4°C$ do not occur. This method of storage is also particularly useful for rare donor blood. The disadvantages are the cost (two to three times higher) and the increased number of manipulations required before use. As a result, frozen cells are little used in the UK at present.

Blood is transfused in cases when the loss has exceeded one litre, and during certain surgical procedures such as cardiovascular surgery when large volumes are required, or for exchange transfusions, for example when there is fetal–maternal blood group incompatability. It is often unnecessary to use whole blood and it may in fact be expedient to use erythrocytes, so avoiding the introduction of the fluid bulk of whole blood. In addition, there is a considerable and growing demand for plasma and platelets, so it is economic to use red cells rather than whole blood when possible. This 'plasma-reduced' blood is obtained by removing 80 per cent of the plasma by centrifugation, and is in common use for post-operative transfusions and in the treatment of some forms of anaemia. The usual collection method for leucocytes is continuous-flow centrifugation. Blood from a donor is passed directly into the centrifuge, a proportion of the leucocytes removed and the remaining leucocyte-depleted blood returned into an arm vein. The harvested cells, enriched in leucocytes, must be used within a short time of collection. A recent development is their use for patients with infections that will not respond to antibiotics, for example in cases of acute leukaemia where infections cause 70 per cent of deaths.

Platelets may be isolated by a similar continuous-flow centrifugation procedure or by serial centrifugation of blood collected in a conventional manner. Platelets are more stable at 22°C than 4°C but have a shelf-life of only a few days. They are employed therapeutically in cases of genetic or acquired platelet deficiency. In leukaemia, when there may be an acquired platelet deficiency due to chemotherapy, platelet transfusion can reduce by 70 per cent the death rate caused by bleeding complications.

Purified plasma proteins or plasma protein fractions are isolated from donated blood or placentas. Some 20 plasma proteins are used therapeutically. However, many more are required for the preparation of specific antibodies, which form the basis of several analytical methods for determining the levels of individual proteins in plasma; currently over 60 different proteins can be assayed by these methods (see p. 154). Such is the extensive use of purified plasma proteins that approximately one million litres of blood are fractionated each year in the USA.

The plasma protein fractions that are most commonly prepared are albumin, γ-globulins, factor VIII, the prothrombin complex (factors II, VII, IX and X) and fibrinogen. These are obtained by the Cohn fractionation, which utilises selective precipitation with ethanol at low temperatures. Albumin is an alternative to plasma or dextran (a high molecular weight polysaccharide) when a blood expander is needed to increase blood volume, as in liver disease or nephritis. Injections of γ-globulins provide passive immunity in cases of severe infections and in antibody-deficiency diseases. Factor VIII, the prothrombin complex and fibrinogen are administered during replacement therapy for absent coagulation factors. The ability to supply factor VIII has extended the life expectancy of haemophiliacs from 16.5 years in 1940 to, currently, a figure little short of that for the rest of the population. Two types of factor VIII preparation are available: cryoprecipitate is a factor VIII-enriched fraction that precipitates from frozen plasma upon thawing; purified factor VIII is obtained by further fractionation of cryoprecipitate. Other proteins being evaluated clinically include plasminogen and antithrombin III as antithrombotic agents, and haptoglobin for use in patients with acute vascular haemolysis.

# Blood Analysis

## Organisation in Hospital Laboratories

Considerable information concerning biochemical changes in the tissues can be obtained from blood analysis. The information is used in diagnosis, prognosis and in following the effect of treatment. A moderate-sized hospital laboratory performs approximately 500,000 tests per annum, this large number reflecting the extent of the information obtainable from blood analysis. The growth in demand for analyses has required the introduction of a high degree of elaborate mechanisation (usually referred to incorrectly as automation) in the medical laboratory. This has been possible because, although the demand for analyses has increased, it has remained for a relatively restricted range of tests. For example, twelve tests usually account for over 80 per cent of the workload of a typical laboratory.

Most hospitals have a biochemistry laboratory which is capable of analysing the majority of the blood samples generated in its area. In a typical hospital 70 per cent of these analyses will be from in-patients, 20 per cent from out-patients and around 5 per cent from general practitioners. Supraregional Assay Service laboratories, usually located in large hospitals, perform the more specialised, and possibly less frequently requested, tests. Such laboratories may specialise in polypeptide hormone, special protein and steroid assays. Hospitals which perform operations like open heart surgery may also have emergency laboratories for blood gas analyses and plasma electrolytes adjacent to the operating theatres.

In addition to the biochemical assays a number of routine haematological parameters are measured. These include the number and size of red cells, their haemoglobin content and the numbers of white cell types; blood clotting times also are determined.

In a routine biochemistry laboratory almost all analyses are performed on mechanised analysis equipment which is capable of automatically performing all the operations which would be performed individually in a manual analysis, for example sampling, pipetting reagent, diluting, mixing, incubation, dialysis, centrifugation, filtration, colorimetry, preparation of standard curves and calculation and printing of results. The operator simply presents the samples to the instrument together with an adequate supply of reagents and regularly checks the performance of the equipment.

There are two types of automatic analyser which are capable of performing the majority of analyses. These sample for a specific assay, either serially (continuous-flow analysers) or in parallel (centrifugal fast analysers). The equipment most commonly used is the *continuous-flow analyser* and the latest models can perform groups of six tests on a single plasma or serum sample. The first group of tests is that for plasma electrolytes ($Na^+$, $K^+$, $Cl^-$ and $HCO_3^-$), urea and creatinine,

Some 36 per cent of all requests are for this group. In many instances these tests are performed to establish whether a patient has some hitherto undetected problem, for example renal failure. The second group of assays includes tests of liver function and calcium metabolism. The plasma is analysed for total protein, albumin, bilirubin, $Ca^{2+}$, phosphate and alkaline phosphatase. They account for some 35 per cent of all requests. A third group of tests is for tissue enzymes (glutamate-oxaloacetate transaminase, lactate dehydrogenase and creatine phosphokinase, for example) the levels of which can indicate the degree of damage to specific tissues. The remaining assays concern those for glucose (15 per cent), and urate and cholesterol, which together account for some 5 per cent.

Assays are now being performed increasingly on *centrifugal fast analysers*. These analysers utilise a multicompartmental rotor in which about 20 samples together with reagents are mixed by centrifugal force. After completion of the reaction the rotor also functions as a multiple spectrophotometer cell. In such an instrument the changes in absorbance in all the samples can be followed continuously and the information processed by computer to give the appropriate printout of information. Because of the speed at which a batch of samples can be analysed, these instruments seem likely to play an increasing role in hospital laboratories in the future.

## Measurement of Inorganic Components

The main cation of plasma is $Na^+$, while $K^+$ and $Ca^{2+}$ are present at concentrations about 2 per cent that of $Na^+$ (Table 10.1). The main anion is $Cl^-$ with lesser amounts of bicarbonate and phosphate. Measurement of $Na^+$, $K^+$, $Cl^-$ and bicarbonate gives an indication of the efficiency of the ion pumps of tissues and of the effectiveness of the kidney as the ultimate controller of the body's ion status. The distribution of ions amongst cells, plasma and interstitial fluid is an important factor in determining water distribution in the body. The measurements of calcium and phosphate are linked together in a bone-function test which permits investigation of parathyroid disorders, for example.

Inorganic ions may be measured by several types of assay — flame photometry, atomic absorption spectroscopy, colorimetry and by using ion-selective electrodes. The $Na^+$ and $K^+$ concentrations are determined simultaneously by flame photometry. Atomic-absorption spectroscopy is used to measure $Ca^{2+}$, $Mg^{2+}$ and sometimes other trace cations (e.g. $Cd^{2+}$, $Cu^{2+}$, $Pb^{2+}$ and $Zn^{2+}$). Colorimetric assays, which involve the reaction of particular reagents with ions to give coloured complexes, are available for the measurement of phosphate, $Ca^{2+}$ and $Cl^-$.

Ion-selective electrodes, which are related to pH electrodes, provide reagentless methods for the sensitive measurement of $Na^+$, $K^+$, $Cl^-$ and $Ca^{2+}$.

## Blood Gas Analysis

Measurements of oxyhaemoglobin concentration and pH are combined with $pCO_2$ and $pO_2$ determinations because all four parameters are interrelated through the Bohr effect and oxygen dissociation curve. The pH value is determined using a conventional pH electrode.

The $CO_2$ electrode is a modified glass pH electrode in which the electrode is bathed in a solution of sodium bicarbonate separated from the blood sample by a gas-permeable membrane. $CO_2$ in the blood diffuses across the membrane and the electrode measures the induced shift in pH.

In the oxygen electrode oxygen is reduced to $H_2O$ at a platinum cathode. By maintaining a constant polarising voltage between anode and cathode, the current flowing is proportional to the oxygen tension. As with the $CO_2$ electrode the oxygen electrode is separated from the blood by a gas-permeable membrane.

The oxyhaemoglobin content of blood is determined spectrophotometrically in a simple instrument such as an oximeter. Both oxy- and de-oxyhaemoglobin absorb light, but their absorption spectra are different. At 850 nm the absorbances of the two forms are identical and, therefore, the absorbance value at this wavelength is a measure of total haemoglobin. In contrast, oxyhaemoglobin absorbs light far more strongly than the deoxy- form at 650 nm. By combining the readings at the two wavelengths the ratio of oxy- to deoxyhaemoglobin can be calculated.

Total haemoglobin and the relative proportions of all four forms — oxyhaemoglobin, deoxyhaemoglobin, methaemoglobin and carboxy-haemoglobin — can also be determined accurately in the laboratory by calculations from absorbance measurements made at four separate wavelengths. Analysis requires a blood sample of 0.4 ml and takes one minute. Knowledge of carboxyhaemoglobin levels is useful in the clinical management of people exposed to carbon monoxide, for example fire victims and smokers, whereas the measurement of methaemoglobin is useful in certain haemoglobinopathies and cases of drug overdose.

## Measurement of Organic Components in Plasma

The organic components of blood that are assayed routinely are urea, creatinine, protein, albumin, bilirubin, glucose, glutamate-oxaloacetate transaminase and alkaline phosphatase. The enzyme assays are discussed in a separate section.

**Table 12.1: Some Enzymes Commonly Assayed as Part of Clinical Diagnosis**

| Enzyme | Organ distribution | Comments |
|---|---|---|
| GOT | Widespread but little in red cells | Analysis of this enzyme started clinical enzymology |
| CPK | Widespread but skeletal muscle is richest source | Used to monitor skeletal and heart muscle disorders, e.g. muscular dystrophy, infarct. Appears in blood after vigorous exercise, hypothermia or surgery |
| γ-GT | Mainly liver | Used as a marker enzyme for hepatocellular disease |
| LDH | Widespread but has distinctive isoenzyme distribution | Used to monitor heart and liver disease |
| Acid phosphatase | Highest specific activity in prostate gland | Used to monitor prostatic cancer |
| Alkaline phosphatase | Easy to assay but widespread in the tissues | Particularly useful in diagnosis of bone diseases where there is increased osteoblastic activity |

*Urea and Creatinine.* Urea, the principal end-product of protein metabolism, is synthesised in the liver and transported in the blood to the kidney for excretion. Elevated blood levels are usually indicative either of kidney disease or increased protein catabolism. Abnormally low levels are rare and often signify severe liver disease. The estimation of urea in blood is based on a non-enzymic reaction in which urea reacts with diacetyl forming a pale yellow compound. The colour is intensified by adding thio-semicarbazide and ferric choride.

Creatinine is an anhydride derived from creatine, a compound particularly characteristic of muscle. Creatine metabolism leads to the synthesis of creatinine which leaks out of the tissue to be removed by the kidneys. The assay is based on the Jaffé reaction whereby creatinine reduces picric acid in alkaline solution to a red compound, picramic acid. Because the rate of creatinine synthesis depends on creatine concentration, which is remarkably constant in muscle, the measurement of plasma creatinine provides an index of muscle mass. Taken in conjunction with the urea value, information is obtained on the rate of protein catabolism; an acceleration of protein catabolism is reflected by an increasing ratio of urea to creatinine.

*Bilirubin.* Circulating bilirubin is predominantly bound, non-covalently, to albumin, whereas low levels of a glucuronic acid conjugate are present in free solution. The concentrations of the two forms may be determined colorimetrically by converting the bilirubin into a red compound, azobilirubin. Measurements performed in the presence or absence of methanol, which releases albumin-bound bilirubin, allow calculation of soluble and total levels, respectively. Abnormally high concentrations of bilirubin in the plasma may be either conjugated or in the free form. Viral hepatitis and cirrhosis are conditions in which conjugation of bilirubin is impaired, thus leading to impaired biliary excretion. The neonatal liver also has a poor capacity to conjugate bilirubin. In haemolytic anaemia accelerated haem degradation may overload an otherwise competent conjugation system and cause accumulation of bilirubin. An obstruction of the bile duct in obstructive jaundice results in raised plasma levels of *both* bilirubin *and* its glucuronide because the conjugate formed in the liver passes back into the blood. Thus a knowledge of the plasma levels of both forms is useful to the clinician when trying to deduce the origins of the jaundice.

*Glucose.* The plasma level of glucose is usually maintained between carefully-controlled limits in order to ensure the efficient operation of tissues particularly dependent on glucose, for example brain and erythrocytes. The rate of glucose production and removal by tissues is controlled by hormones so disturbances in blood glucose are often symptomatic of disorders of hormone production. For example, hyperglycaemia is usually associated with diabetes mellitus in which there may be an inadequate amount of insulin or an inability to respond to the insulin that is present. In contrast, adrenal cortex insufficiency, Addison's disease, leads to hypoglycaemia because the glucocorticoids required to maintain gluconeogenesis are not produced.

The specific assay for glucose is based on the reaction catalysed by glucose oxidase. In this reaction molecular oxygen is reduced to hydrogen peroxide which in turn oxidises a dye with the formation of a coloured complex.

*Plasma Proteins.* The analysis of the complex mixture of about 100 proteins in plasma is performed at several levels of discrimination and refinement according to the purpose of the investigation. The initial, simple routine analyses are for *total protein* and *albumin*, and may be accompanied by a request for an electrophoretic separation of the proteins. In contrast, the assay of the various individual proteins has become a province of analysis in its own right and is accorded a separate section in this chapter.

*Albumin* is the major protein in plasma and has the ability to bind a wide range of natural and synthetic compounds (see Chapter 11). It is this latter property that is exploited in the assay of albumin. Dyes, for example, bromocresol green, bind strongly to albumin and the coloured complex can be measured colorimetrically. The total protein content, approximately 70 $gl^{-1}$, is estimated by the Biuret reaction which relies on the formation of a red complex between peptide bonds and copper ions in alkaline solution. A reduction in the amount of albumin relative to total protein can indicate either: (a) accelerated catabolism, as occurs during infection; (b) decreased synthesis caused by insufficient protein intake or impaired synthesis in cirrhosis of the liver; or (c) excessive loss of protein. Nephrotic syndrome is an example of a disorder where albumin is lost selectively. The kidneys are unable to retain proteins and, because the filtration is based mainly on molecular size, this affects particularly the smaller plasma proteins such as albumin and to a lesser extent IgG.

The most useful analytical system would effect a complete separation of all proteins in plasma; in practice, it is possible only to resolve them into groups of proteins. The most widely used system of classifying plasma proteins is by their electrophoretic mobility at pH 8.6. If this separation is performed on a high resolution support medium, such as cellulose acetate or agarose, the serum sample is resolved into a series of bands which, in order of decreasing mobility towards the anode, are designated albumin, α-globulins, β-globulins and γ-globulins. α and β groups can be divided into $\alpha_1$, $\alpha_2$ and $\beta_1$, $\beta_2$ subgroups. *Electrophoretic analysis* of plasma reveals an extra component migrating between the β and γ bands: this is fibrinogen. To detect the presence of these protein bands the electrophoretic strip is stained with a dye that reacts with all proteins and by measuring the intensity of the stain a quantitative estimation of each band's contribution of the total is obtained.

Although a disease may often be accompanied by a change in the concentration of a plasma protein, because most of the bands observed are in fact mixtures of proteins, only if one of the major components changes in concentration will an alteration be apparent. This limits the value of electrophoretic analysis to those conditions where gross changes do occur. For example, with progressive liver failure the levels of hepatic-synthesised proteins decline; this includes virtually all plasma proteins except immunoglobulins. Inflammation, wherever it occurs in the body, is accompanied by an increase in the synthesis by the liver of 'acute-phase proteins' (fibrinogen, $\alpha_1$-acid protease and $\alpha_1$-acid glycoprotein). The $\gamma$-globulin fraction, which contains predominantly immunoglobulins, is elevated in hypergammaglobulinaemia as a consequence of the overproduction of antibody by an uncontrolled clone of plasma cells. Another condition detectable by electrophoretic analysis is nephrotic syndrome. Here the loss of albumin in the urine results in albumin being a very weak minor band, whereas normally this plasma protein accounts for approximately half of the total protein.

To measure particular proteins in plasma advantage is taken of the specificity inherent in an antibody reacting with its antigen. By quantitating the formation of the antibody–antigen complexes a measure is obtained of the amount of antigen, that is plasma protein, present in the sample. There are a number of *immunological techniques* widely used in clinical laboratories and all these procedures are based on the binding of the antibody and antigen; the differences arise in the way in which the process is quantified.

The antibodies are raised by injecting the pure human protein into an animal, for example goat or sheep, at intervals over a period of several weeks. The foreign protein stimulates the animal's immune system so that the animal's serum contains antibodies directed against the human protein.

*Immunoelectrophoresis* combines the separation capabilities of electrophoresis with the specificity of the antibody–antigen reaction. The first stage is serum electrophoresis in agarose. This is followed by an immunodiffusion step in which an antiserum is allowed to diffuse into the gel from a channel alongside the electrophoretic track. As the serum proteins and antibodies diffuse through the gel, immune complexes form that precipitate as arcs (Figure 12.1). Immunoelectrophoresis of human serum using an antiserum directed against all human serum proteins produces a pattern of about 30 intersecting precipitin arcs corresponding to the major proteins. The technique is useful for detecting particular proteins or those with an altered mobility but its value is limited because it is not quantitative.

Another immunological analytical technique, known as *rocket electrophoresis*, exploits the fact that, during electrophoresis, antibodies ($\gamma$-globulin fraction) migrate slowly, if at all, in comparison with the antigenic proteins. The sample is subjected to electrophoresis in an agarose gel medium impregnated with antiserum raised against the particular protein to be analysed. In the electric field the serum proteins migrate into the antibody-containing gel. Initially the protein to be assayed forms relatively small antigen–antibody complexes which continue to migrate because of the charge on the antigen. However, the complex grows on encountering further antibody molecules and eventually precipitates. Visualisation of the immunoprecipitate shows that it is in the form of a pair of precipitin lines converging to a peak. The peak shape resembles the outline of a rocket, hence the technique has the trivial name of rocket electrophoresis. The height of this rocket is proportional to the amount of antigen. The advantage of rocket electrophoresis is speed: a complete quantitative analysis is possible in a few hours. Typical serum proteins assayed by rocket electrophoresis are caeruloplasmin, $\alpha_1$-acid protease, haptoglobin, transferrin, the C3 component of complement and antithrombin III.

**Figure 12.1: Immunoelectrophoresis: (a) Electrophoresis of Serum Sample; (b) Diffusion of Antiserum and Serum Proteins Through Agarose Gel; (c) Formation of Precipitin Arcs**

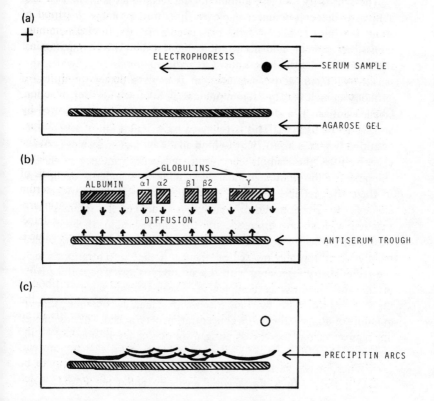

The immunoanalytical procedures described so far require gel media. Sometimes this medium is necessary to control the rate of diffusion but more often the presence of the gel prevents disturbances to gradients during the extended periods required for the development of immune complexes. Antibodies and antigens will react in free solution *in vitro*, but the complexes develop slowly and consume large amounts of antiserum. Complex formation can be accelerated by the addition of a polymer such as polyethylene glycol, which has the added advantage of decreasing the solubility of the antigen–antibody complexes. The resultant large aggregates scatter light far more than the individual proteins and thus the degree of scattering is a measure of the extent of the antigen–antibody reaction and hence the quantity of antigen. The measurement of turbidity, known as *nephelometry*, has the advantage to a clinical laboratory of lending itself readily to automatic analysis. It is one of the commonest methods employed for quantitation of the various immunoglobulin classes in serum.

The sensitivity of these immunological techniques is limited by the ability to detect immune complexes. For most proteins this limit is about 0.5 mg $l^{-1}$. Many of the plasma constituents, for example hormones, are present at concentrations far less than this. Estimation of these requires a technique known as *radioimmunoassay* in which the sensitivity of radioactive measurement is used to detect the antigen–antibody complexes. In radioimmunoassay a known amount of radiolabelled antigen is added to a serum sample containing the unlabelled antigen to be assayed. This is followed by the addition of an amount of antibody that is insufficient to bind all the antigen molecules present. Consequently the radiolabelled and unlabelled antigen molecules compete for the limited number of binding sites and the occupancy of these sites reflects the ratio of labelled to unlabelled antigen. For example, to assay insulin $^{125}$I-insulin is added to the serum sample together with an antiserum specific for human insulin. If the number of insulin molecules in the serum is small the binding sites are occupied mainly by $^{125}$I-insulin molecules whereas if the concentration is high the radioactivity associated with the antibody is low. The radioactivity in the antigen–antibody complex is measured after separation from unreacted antigen and, by comparison with standards containing known amounts of antigen, the serum concentration determined. The complexes are removed from the excess antigen by one of several methods. One common procedure is to add a second antibody that is specific for the first. Thus, if the antibody directed against human insulin had been raised in sheep, then the second antibody could be an anti-sheep antiserum obtained, for instance, from goats by injecting them with sheep antibodies. The second antibody combines with the first to form an insoluble complex which is removed by filtration or centrifugation. An alternative technique is to covalently link the antibody to an insoluble polymer, for example, cellulose fibres or polysaccharide beads. After the antigen-antibody reaction is complete, the polymer and associated complexes are removed and the radioactivity measured.

Radioimmunoassay has been applied successfully to the analysis of a range of polypeptide hormones including insulin, parathyroid hormone, growth hormone, FSH, LH and prolactin. The assay of $\alpha$-fetoprotein in maternal blood during pregnancy is another important application. Elevated levels of this protein are associated with the fetal neural tube defects spina bifida and anencephaly.

Radioimmunoassay can be adapted for the analysis of a wide range of non-protein blood components. Although these components may be non-antigenic, advantage is taken of the hapten reaction in which the small molecule is coupled to a larger protein. This complex is antigenic and causes the animal to produce antibodies directed against a range of antigenic determinants, including the hapten. Using this hapten reaction specific antibodies can be raised against the thyroid hormones, $T_3$ and $T_4$, and all the major steroid hormones.

Although the radioimmunoassay procedure described here depends on the roles of antibodies as specific binding agents, any protein that recognises and binds tightly to a particular species of molecule can be used in the same way. For example, a folate-binding protein obtained from pig serum is used to assay serum folate. Similarly pig stomach intrinsic factor is used to estimate vitamin $B_{12}$ concentrations.

The principles described for radioimmunoassay have wider applications. Instead of achieving sensitivity through the use of radioisotopes, other less hazardous ligands can be coupled to the antibody to provide the means of detecting the formation of antibody-antigen complexes. For example, a fluorescent compound covalently linked to the antibody, although not as sensitive as a radiolabel, provides adequate limits of detection for many antigens. Recent developments have been directed to coupling an enzyme (e.g. alkaline phosphatase) to the antibody. After separation of antigen-antibody-enzyme complexes from uncomplexed antibody-enzyme, the residual enzyme activity is measured. Because the enzyme generates so many molecules of product, very small amounts of antigen can be detected.

An important point concerning all immunological assays is that the antibody recognises an antigenically-reactive protein and does not measure its biological activity. For example, a patient with haemophilia and having a defective factor VIII activity may have an apparently normal level of the protein as estimated by immunoassay, but biologically the protein is non-functional.

*Assay of Tissue Enzymes in Plasma*

When organs are damaged part of their enzyme complement is released into the plasma. In a healthy person the levels of intracellular enzymes in plasma are very low and result from cellular turnover. Compared with plasma, the tissues contain $10^3$-$10^4$ times higher concentrations of soluble enzymes within their cells, so even slight damage to a large organ may result in the release of measurable quantities of enzymes. Intracellular enzymes released into the plasma are inactivated and removed within a few days. Consequently, by monitoring plasma enzyme levels the time-course of a disease can be followed. The amount of an enzyme released depends on the concentration of that enzyme in the organ and the extent of tissue damage. Knowledge of the cellular location of the enzyme provides useful information also. For instance, in a reversible inflammatory disease, such as acute hepatitis, only cytoplasmic enzymes appear in plasma. This is because ATP depletion causes permeability changes in the cellular plasma membrane. Chronic conditions, especially those involving ischaemic damage and necrosis, for example cirrhosis, lead to enzyme release from cell organelles. However, there are exceptions to these simple guidelines and accurate interpretation of plasma enzyme levels in disease diagnosis requires considerable experience.

Theoretically damage to any organ can be assessed by enzyme analysis. However, in practice enzyme assays are most useful in detecting damage to the liver, heart, skeletal muscle, blood cells, bone, pancreas and prostate. The cells of each tissue have a characteristic complement of enzymes – their 'enzyme pattern'. This pattern may be mirrored in the plasma if there is massive tissue damage. For example, the cardiac cell 'enzyme pattern' appears in plasma after a severe myocardial infarct. Further, certain enzymes are confined to one or two tissues, for example creatine phosphokinase (CPK) is characteristic of striated muscle and $\gamma$-glutamyltransferase ($\gamma$-GT) of the liver (see Table 12.1). In contrast, other enzymes occur in many tissues, for example lactate dehydrogenase (LDH) and glutamate-oxaloacetate transaminase (GOT). In these cases it may be possible to identify isoenzyme forms originating from different tissues (see below).

The main criteria in designing a routine enzyme assay are reliability, reproducibility and simplicity. When large numbers of samples are to be assayed it is also an advantage if the assay can be mechanised and is inexpensive. Assay conditions must be selected to ensure that the rate of reaction is proportional to enzyme concentration. Generally this means that the enzyme concentration must be low and the substrate concentration sufficiently high to ensure enzyme saturation. Other parameters, such as cofactor concentration, pH, buffer composition and temperature, must be standardised also. When suitable assay conditions are established the enzyme activity can be calculated from the rate of product formation and related to the size of the blood sample. The blood cells themselves contain relatively large amounts of enzyme and so blood collection and subsequent handling must be carried out carefully to avoid haemolysis. For instance, release of acid phosphatase from erythrocytes can complicate the assay of prostatic acid phosphatase in the diagnosis of prostatic cancer (see below). Other factors may also complicate the interpretation of enzyme assay data; these factors include individual variation due to age, sex and general physical well-being.

As an example of the use of enzyme assays in diagnosis some aspects of liver disease will be considered. In certain cases enzyme assays may provide an indication of disease before there are obvious clinical manifestations. For example, in acute viral hepatitis, levels of GOT and $\gamma$-GT frequently increase significantly several days before the ability to excrete bilirubin is impaired and the patient becomes jaundiced (Figure 12.2). This can be important in prophylaxis during epidemics. Unlike GOT and $\gamma$-GT, the plasma level of acid phosphatase rises only slightly during acute viral hepatitis (Figure 12.2) unless it is associated with cholestasis. Biliary obstruction also leads to a release of $\gamma$-GT over an extended period and a slower rate of return of GOT to its normal level. It may be possible to distinguish between intrahepatic cholestasis and extrahepatic obstructive jaundice because the latter condition often leads to a release of glutamate dehydrogenase (GDH). In other respects the pattern of enzyme release is very similar. Thus, by comparing the relative amounts of different enzymes released, it is often possible to obtain good indications of the nature of liver disease.

Although enzyme assays are used as part of initial diagnosis, they are probably most valuable for monitoring the progress of disease and the efficacy of therapy. Once the enzyme levels are known for a particular patient, even small changes are significant indicators to the clinician, and may also reveal the involvement of other organs. For instance, liver congestion often occurs in bed-ridden patients a few days after damage to the heart, as in myocardial infarct (see Figure 12.3). The enzymes released initially are those characteristic of heart muscle, that is CPK and the LDH$_1$ isoenzyme of lactate dehydrogenase, and the amounts indicate the severity of the infarct. If no further damage occurs, the levels return quickly to normal. However, if the liver becomes congested, due to inefficient pumping by the right side of the heart, there will be a further release of enzymes. Furthermore, these enzymes will be characteristic of the liver, that is $\gamma$-GT and the LDH$_5$ isoenzyme of lactate dehydrogenase, and thus the involvement of the liver in non-hepatic disease can be detected.

**Figure 12.2: Enzyme and Bilirubin Levels in Plasma During Acute Viral Hepatitis**

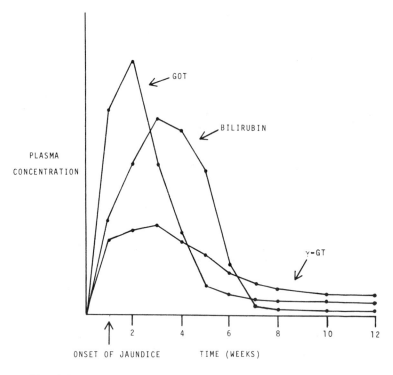

The ability to recognise the different isoenzyme forms of LDH plays an essential part in monitoring the release of LDH activity from different tissues. The isoenzymes can be separated and recognised by gel electrophoresis by exploiting differences in charge. However, for routine analysis this method is complex and cumbersome. A simpler procedure exploits differences in thermal stability. $LDH_5$ is denatured by heating the plasma at $65°C$ for ten minutes whereas $LDH_1$ is not. Therefore, LDH activity assayed before and after heating provides a means of determining the relative proportions of the two isoenzymes. Another method makes use of the fact that $LDH_1$, but not $LDH_5$, is able to oxidise α-hydroxybutyrate as an alternative substrate to lactate.

**Figure 12.3: Plasma Enzyme Levels After Myocardial Infarct Followed by Liver Congestion**

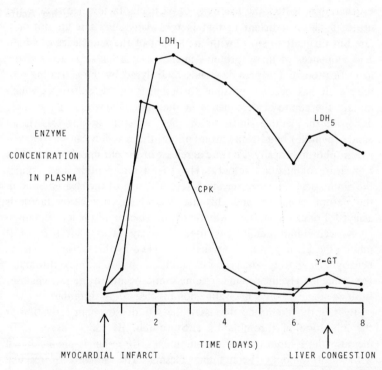

Another example of using differential isoenzyme analysis is the measurement of acid phosphatase in the diagnosis of prostatic cancer. A large proportion of males eventually suffer from some form of prostatic infection or disease and prostatic cancer is quite common. Most patients with metastatic prostatic carcinoma have elevated plasma acid phosphatase levels which may be measured in the presence of acid phosphatase released from other tissues by using specific substrates or inhibitors for the prostatic isoenzyme. A recent advance is the development of a specific radioimmunoassay (see above) for the prostatic acid phosphatase. This provides a more sensitive and reliable measure of the prostatic isoenzyme and increases the feasibility of diagnosis of prostatic carcinoma before metastases have formed.

## Analysis of Components of the Coagulation System

The routine method of measuring coagulation is the rate of clot (fibrin) formation. If the sequence of reactions is broken, or an inhibitor is present, then the rate is retarded. Because so many factors are involved, the assay is extremely difficult to control and reproduce accurately. Furthermore, to test the reactivity of particular factors necessitates the use of a serum deficient in that factor. Many such sera are not only rare but difficult to store. With the advent of the knowledge of amino acid sequences of many proteins in the coagulation sequence and the identification of the peptide bonds hydrolysed by the activating proteinase, it has become possible to design synthetic substrates which imitate the amino acid sequence in the natural substrate. By suitable design of the peptide substrate and by covalently linking this via an amide (peptide) bond to a chromophore, the reaction can be followed spectrophotometrically. The rate of release of the chromophore provides a measure of enzyme activity. The tri- or tetrapeptide used mimics the amino acid sequence on the N-terminal side of the cleaved bond in the natural substrate and this dictates which coagulation factor is assayed. For example, the synthetic peptide benzoyl-isoleucyl-glutamyl-glycyl-arginyl-$p$-nitroanilide is used to assay factor Xa, because it mimics the site in prothrombin that is attacked by this factor. Similarly, benzoyl-glycyl-prolyl-arginyl-$p$-nitroanilide is hydrolised by thrombin: the sequence glycine-proline-arginine-valine occurs in the (A) chain of fibrinogen and the argenine-valine bond is cleaved by thrombin.

By slightly modifying the assay the rate of thrombin activation or the inhibition of thrombin by antithrombin III can be assayed. To measure the inhibitor a known amount of thrombin is pre-incubated with antithrombin III-containing plasma. The amount of residual thrombin is inversely proportional to the amount of antithrombin III. Such assays are used to monitor oestrogen or anticoagulant therapy.

## Blood Typing

If a person receives incompatible blood by transfusion, haemolysis usually occurs and death may result. Haemolysis can be very rapid with 450 ml of blood being haemolysed in one to two hours, partly

Figure 12.4: Agglutination Assays for Determining Blood Group Antigens and Antibodies: (a) Agglutination by 'Complete' Antibodies; (b) Binding of 'Incomplete' Antibodies; (c) Use of Anti-human IgG Antibodies to Complete the Agglutination Reaction

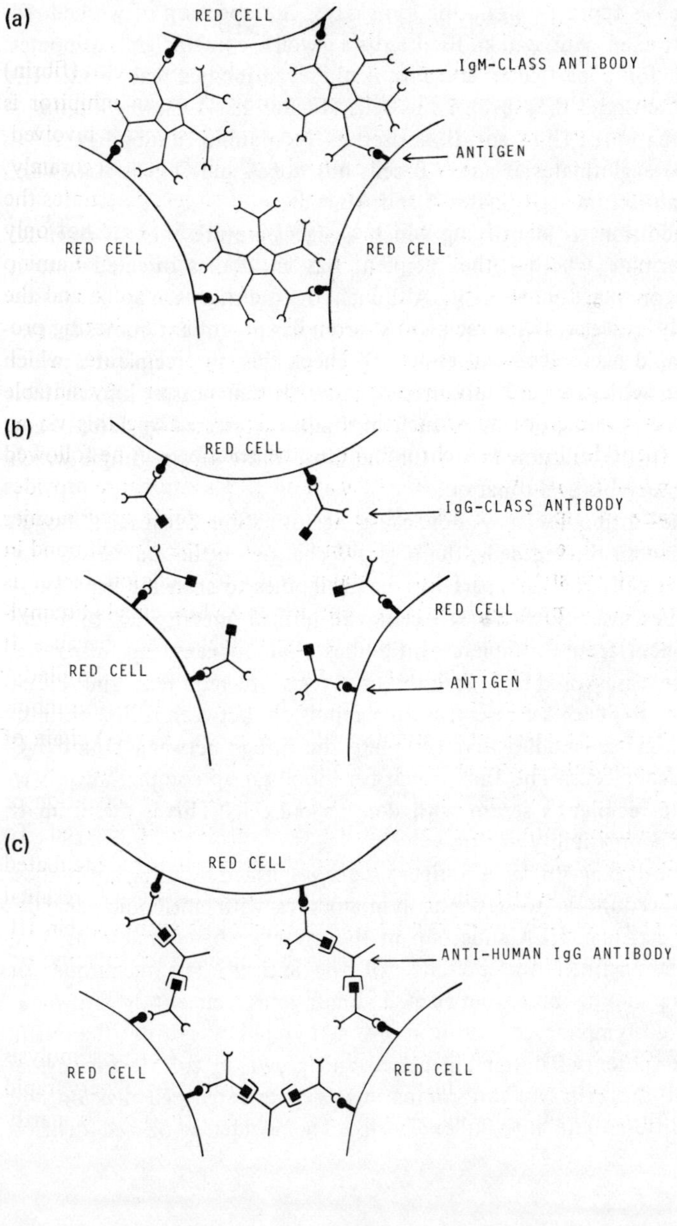

(a)

(b)

(c)

in the circulation and partly in the liver and spleen. The extent and speed of haemolysis depend upon the potency of red cell antigens and the reactivity of the recipient's antibodies. In practice, the only antigens sufficiently potent to present a clinical problem are those within the ABO system and the Rhesus D antigen (see Chapter 8).

Blood is typed by agglutination assays. A suspension of washed red cells is mixed with standardised antisera which contain IgM antibodies specific for a particular antigen. If these antibodies react with cell-surface antigens they bridge the cells (Figure 12.4) causing clumping (agglutination). Thus anti-B antiserum (containing antibodies to B antigen) agglutinates B and AB cells but not A and O cells. Likewise, anti-A antiserum agglutinates A and AB cells.

In addition to identifying which antigens are present it is essential to determine whether the recipient has antibodies directed against antigen on the donor's cells. Although these antigens may be classed normally as weak, if the recipient's serum has potent antibodies against them rapid haemolysis will ensue. To check this, the recipient's serum is tested with standard mixtures of red cells that possess known antigens. The potency or titre of such antibodies is assessed by determining the reciprocal dilution factor for the most dilute serum sample which gives observable agglutination.

Some antibodies may not cause agglutination because, although they bind to the red cells, the negative charges on the red cell surface hold the cells too far apart for the antibodies to span the gap. These antibodies, usually of the IgG class, are termed 'incomplete' to distinguish them from 'complete' antibodies that do cause agglutination. In these cases modified agglutination tests are required using either albumin to decrease electrostatic repulsion between cells or anti-human IgG antibodies that complete the bridge between bound IgG on adjacent cells. The final check for blood group compatibility is to incubate recipient's serum with donor's red cells. This is the ultimate test for compatibility *in vitro*.

Tissue typing for HLA antigens is accomplished by using antibodies. The procedure is to test the lymphocytes with antibodies specific for a particular HLA subgroup in the presence of complement. Cell lysis demonstrates the presence of the antigen. The technique for assessing compatibility between a donor and recipient is known as the mixed lymphocyte culture reaction. Lymphocytes from the recipient and potential donor are mixed and grown in culture for several days. If the cells are not compatible they undergo cell division and some differentiate into killer T-cells. The incidence of cell division is monitored by assaying the incorporation of radioisotopically-labelled thymidine into DNA.

## Further Reading

Allen, P.C., Hill, E.A. and Stokes, A.M. (1977) *Plasma Proteins, Analytical and Preparative Techniques*, Blackwell, Oxford

Gunson, H.H. and Dunleavy, J. (1982) 'Modern Trends in Blood Transfusion: The Use of Computerization' in A.V. Hoffbrand (ed.), *Recent Advances in Hematology, 3*, Ch. 15, Churchill Livingstone, Edinburgh

Moss, D.W. (1980) 'Clinical Enzymology – A Perspective', *Enzyme, 25*, 2-12

Schwick, H.G. and Heide, K. (1977) 'Trends in Human Plasma Protein Research', *Trends in Biochemical Sciences, 2*, 125-8

Sykes, M.K., Vickers, M.D., Hull, C.J. and Winterburn, P.J. (1981) *Principles of Clinical Measurements*, 2nd edn, Blackwell Scientific Publications, Oxford

# 13 PERSPECTIVES

Blood is the most accessible of the body fluids and there is little doubt that this accounts for the quite spectacular progress in the field of blood biochemistry. It is also important to recognise that research into blood biochemistry has led to the discovery of a number of fundamental aspects of biochemistry and molecular biology. For example, the recognition and study of abnormal haemoglobins, together with investigations of their causes, has given much valuable information on both structure–function relationships in proteins and the expression of genetic information. In addition, the study of globin synthesis in cell-free preparations from reticulocytes has provided important information on the mechanism of protein biosynthesis in eukaryotes.

The realistic expectation is that current and future experiments will provide further insights into many aspects of biochemistry in general and, perhaps more importantly, on the understanding of disease processes, their clinical management, prevention and cure. This final chapter pinpoints some aspects of the biochemistry of blood in which particular advances have been made recently, or where breakthroughs in certain techniques provide promise of rapid developments in the future. It does not aim to be comprehensive, but rather seeks to stimulate interest in the subject as a whole in the hope that some of our readers might themselves pursue such aspects of medical research.

One of the growth points in biochemistry is that of genetics. In particular, research into the molecular basis of several genetical phenomena, such as the structure and expression of genes, is a rapidly expanding field. Using genetic engineering techniques, it has been demonstrated that several mammalian genes, including those of antibodies and globin, contain intervening, non-translated sequences (introns). This discovery has not only stimulated research in this area but has also caused a major rethink about the way in which information stored in DNA is transcribed and translated into proteins. Thus, it has recently become clear that the primary transcription product of the gene is not mRNA, but a longer RNA precursor containing the introns that are subsequently excised before the true mRNA is translated by the ribosomes. There has been much speculation about the function of introns. The gene for the immunoglobulin heavy chain has three introns and four exons (i.e. expressed sequences) and the exons code for the four functional

168

domains of the antibody molecule. Thus, there seems to be a direct relationship between the organisation of DNA within the gene and functional organisation within the protein.

In the case of haemoglobin, such a clear relationship does not exist. Detailed studies have been made of the three-dimensional structure of haemoglobin and its many variants and certain structure–function relationships are discernible. Five of these are recognisable: they comprise residues in contact with the haem, those involved in the $\alpha_1\beta_1$ or $\alpha_1\beta_2$ contacts, residues in the $\beta$ subunit involved in 2,3-DPG binding and those participating in the Bohr effect (see Chapter 3). The $\beta$-globin gene contains two introns and three exons; the exons correspond fairly well with the regions involved in haem, $\alpha_1\beta$, and $\alpha_1\beta_2$ contacts. But there is no such clear assignment for those residues involved in 2,3-DPG binding or the Bohr effect. A possible explanation may be that these regulatory devices developed after the general structure of haemoglobin and its genes had evolved, and that they have been accommodated without changes in the basic structure–function relationships of the $\beta$-globin gene and its translation product.

Studies designed to determine gene structure involve the technique of gene cloning and it is now possible using recombinant DNA technology to prepare $\alpha$- and $\beta$-globin genes in amounts sufficient for sequence analysis. This technique will facilitate a more detailed analysis, at the genetic level, of some of the many haemoglobin variants; the information gained should in turn shed light on the precise defects in, for example, the $\beta$-thalassaemias. The $\beta+$ thalassaemia genes of two patients have been cloned successfully, but as yet the molecular basis of the defects in these genes has not been discovered. Analysis of the structure of the $\beta$-globin gene in cells cultured from the amniotic fluid of a mother suspected of carrying a child with a $\beta$-thalassaemia (due to a deletion in the $\beta$-globin gene) has already been used for the prenatal diagnosis of this disease.

The mechanism by which $\gamma$-globin (present in fetal haemoglobin) synthesis is turned off is not understood; analysis of the $\gamma$- and $\delta$-globin genes, and that part of the chromosome between them, might aid in solving this problem. The clinical significance of this work is clear: if $\gamma$-chain synthesis could be reactivated it would be possible to control $\beta$-thalassaemia, sickle cell anaemia and other haemoglobinopathies by reverting to the production of functional fetal haemoglobin synthesis.

Another approach to the problem of defective haemoglobin that is being pursued is the insertion of a correctly functioning globin gene into erythropoietic cells. This is achieved by growing a sample of the patient's erythropoietic cells in the laboratory and inserting into them the appropriate globin gene. Samples of these cells are transplanted back into the bone marrow of the patient. This approach must be paralleled by the development of improved culture techniques for erythropoietic stem cells. Already there has been some progress in this direction and replacement therapy has been achieved in two cases of β-thalassaemia. However, there is some debate as to whether the time is yet ripe for this type of experiment in humans, especially since it is fully appreciated that there is still much to learn about the structure, expression and control of genes.

An alternative therapy for sickle cell anaemia is the use of compounds that increase the affinity of haemoglobin for oxygen by stabilising the $R$-state. Likewise, compounds that stabilise the $T$-state could be used for conditions in which oxygen delivery to the tissues is impaired by a haemoglobin with a high affinity for oxygen. There are several examples of this including many abnormal haemoglobins (e.g. see Table 3.3), the early anaemia of premature babies, and methaemoglobin and carboxy-haemoglobin anaemias. A further application of such compounds is in rapidly corrected acidosis, for instance, acute diabetic acidosis, when a low $P_{50}$ value cannot be compensated, because insufficient 2,3-DPG is available. Compounds that stabilise the $T$-state were designed originally as analogues of 2,3-DPG, and further development of these compounds and those that stabilise the $R$-state is continuing.

Another area of genetics in which several important advances have been made in recent years is that concerned with the genetic basis of the immune system. The significance of research in this field is demonstrated by the award of the 1980 Nobel Prize for medicine to Snell, Dausset and Benacerraf. Between them they have pioneered investigations leading to the discovery of the HLA system and the immune response (Ir) genes, which code for the HLA histocompatibility antigens present on the surface of cells (see Chapter 5). Because of this work the success rate of transplant operations has been improved and tissues are now typed for their histocompatibility antigens before being used in transplant operations. Certain HLA types are linked to specific diseases that have an immunological basis or involvement. This link may be provided by the T cells, whose functions are controlled by Ir genes. Thus, it seems that as more is learnt about the genetic organisation of the immune-response system, we shall not only understand better the functions of, for instance, the T cells, but also the basis of diseases like rheumatoid arthritis involving (auto)immune phenomena.

In Chapter 5 the technique of fusing antibody-producing cells with myeloma cells to produce hybrid cells which synthesise monoclonal antibodies is described. These hybrid cells can be used either directly to produce extremely pure antibody or they can be injected into mice where they form tumours which produce larger amounts of less pure (~ 95 per cent) antibody. Monoclonal antibodies against a wide range of antigens (e.g. histocompatibility antigens, cell surface components, hormones, enzymes, red cells and viruses) have been prepared in this way. Although the technique is very new and requires considerable further development, it is already clear that there is a wide spectrum of medical applications for monoclonal antibodies. These applications include blood and tissue typing, radioimmunoassay and the identification of bacteria and viruses. The technique could also be exploited to increase the yield of interferon for use in antiviral or anti-cancer therapy.

The most common cause of death in the Western World is vascular disease, and for this reason advances in this branch of medicine could well have a significant impact on life expectancy. Environmental factors have emerged as the more important influences responsible for the development of thrombosis. Thus, theoretically, the incidence of thrombosis could be reduced by changes in life-style and diet. Such a change seems unlikely at present and thus there remains the necessity for anti-thrombotic drugs.

The major features of the haemostatic systems and the way in which they are controlled now seem clear (see Chapter 9), and the mechanisms of action of the more commonly-used anti-thrombotic agents are established. This knowledge may well allow the design of additional anti-thrombotic drugs with specific modes of action. For example, selective inhibitors of the enzymes responsible for the synthesis of specific platelet prostaglandins could be effective in the prevention of arterial thrombosis. The use of urokinase, a very promising anti-thrombotic agent, is restricted by its high cost, but if the urokinase gene could be isolated and cloned in a bacterium it might well allow cheaper production of urokinase and thus its use on an extended scale.

## Further Reading

Dausset, J. (1981) 'The Major Histocompatibility Complex in Man. Past, Present and Future Concepts', *Science, 213*, 1469–74

Edwards, P.A.W. (1981) 'Some Properties and Applications of Monoclonal Antibodies', *Biochemical Journal, 200*, 1–10

Kan, Y.W. and Dozy, A.M. (1978) 'Antenatal Diagnosis of Sickle-cell Anaemia by DNA Analysis of Amniotic – fluid Cells', *Lancet, II*, 910–11

Lewin, R. (1981) 'Evolutionary History Written in Globin Genes', *Science, 214*, 426-9

Nathan, D.G. (1979) 'Progress in Thalassemia Research', *Nature, 280*, 275-6

Samson, D., Chanarin, I. and Reid, C.D.L. (1981) 'Eecent Advances in Haematology', *Postgraduate Medical Journal, 57*, 139-49

Wade, N. (1981) 'Gene Therapy Caught in More Entanglements', *Science, 212*, 24-5

# INDEX

21